Monte Carlo Methods in
Mechanics of Fluid and Gas

Monte Carlo Methods in Mechanics of Fluid and Gas

O. M. Belotserkovskii
Russian Academy of Sciences, Russia

Y. I. Khlopkov
Russian Academy of Natural Sciences, Russia

 **World Scientific**

NEW JERSEY · LONDON · SINGAPORE · BEIJING · SHANGHAI · HONG KONG · TAIPEI · CHENNAI

Published by

World Scientific Publishing Co. Pte. Ltd.

5 Toh Tuck Link, Singapore 596224

USA office: 27 Warren Street, Suite 401-402, Hackensack, NJ 07601

UK office: 57 Shelton Street, Covent Garden, London WC2H 9HE

British Library Cataloguing-in-Publication Data
A catalogue record for this book is available from the British Library.

MONTE CARLO METHODS IN MECHANICS OF FLUID AND GAS

ISBN-13 978-981-4282-35-2
ISBN-10 981-4282-35-9

Typeset by Stallion Press
Email: enquiries@stallionpress.com

Printed in Singapore.

Preface

Our dedication to the pioneers of the use of Monte Carlo methods in mechanics of fluid and gas in Russia Vladimir Alexandrovich Perepukhov and Vitaliy Evgenjevich Yanitskii.

The beginning of the third millennium is characterized by the global uniqueness of the human civilization. The possibilities of humanity in energetic properties of the industrial processes and of the armament systems became to be comparable with similar properties of the natural processes. It concerns even such energy-consuming processes, as the natural cataclysms. On the one hand, this fact appears as the evidence of the general progress in the development of humanity. On the other hand, this peculiarity evokes a serious misgiving, since it is threatening just the possibility of the further existence of a human civilization. And such a misgiving is connected not only with a possibility of the global thermonuclear wars with unpredictable consequences, but also with the everyday activity on the security of a public life. For example, one of the important factors is the hypothetical global state of climate of Earth. It is assumed that in the result of the large quantities of smoke and soot, which are carried out into the stratosphere through the spacious fires by the explosion of 30–40% of nuclear war-charges, accumulated in the world, the temperature throughout the whole planet will be lowered down to the Arctic values, as a result of the essential increase of the quantity of reflected solar rays. The possibility of appearance of a nuclear winter was forecasted by Charles Sagan in the USA and confirmed in Russia by the computations of V.V. Alexandrov.

The everyday activity on the security of a public life leads both to the accidental large-scale ecological catastrophes, and to the gradually accumulating pollution of the environment (V.P. Dymnikov). Considered in the present monograph are some fundamental problems connected with these subjects. Presented here are the statistical methods of mathematical modeling for various models of the flow of fluid and gas, within the wide range of the characteristical parameters. The models of flow are ranging from the hypersonic flows of strongly rarefied gases (gaseous flows near the Earth's satellites at the orbits and near the apparatuses descending from the orbit), which are influencing the ecological state of the nearest space, and up to the turbulent flows modeling both the atmospheric phenomena and the processes of flow about the modern flying machines. Described are the modern effective numerical methods, developed both by the authors themselves and by other specialists and intended for the computer realization of these models. The problems considered belong to the classes of three-dimensional evolutional problems, based on the equations of mathematical physics, for the overwhelming majority of which are not proved even such a fundamental mathematical motions, as the theorems of existence and uniquity, even in the considerably simplified formulations. The study of such problems, at the present stage of the development of science, when the traditional analytical methods of investigation have, in a certain degree, exhausted themselves, is carried out, mainly, with the help of a computational experiment.

The revelation of the methods of statistical modeling (Monte Carlo) in various areas of the applied mathematics is connected, as a rule, with the necessity of solution of the qualitatively new problems, arising from the needs of practice. Such a situation appeared by the creation of the atomic weapon, at the initial stage of a mastering of space, by the investigation of the phenomena of atmospheric optics, of the physical chemistry, and of the modeling of turbulence (G. von Neumann, Metropolis N., Unlam S., Vladimirov V.S., Sobol I.M., MArchuk G.I., Ermakov S.M., Mikhailov G.A., Bird G.A., Haviland J.K., Lavin M.D., Pullin D.I., Kogan M.N., Perepukhov V.A., Beloserkovskii O.M., Yanitskii V.E., Ivanov M.S., and Eropheev A.I.).

As one of the more or less successful definitions of the Monte Carlo methods, it would be possible to present the following one:

The Monte Carlo methods present in themselves the numerical methods of solution of the mathematical problems (sets of the algebraic, differential, or integral equations) and the direct statistical modeling of the processes (physical, chemical, biological, economical, and social) with the help of obtainment (generation) of the accidental numbers and the transformation of those.

The book contains, in the reasonable proportions, those formulations and solutions, which already proved to be classical ones, as well as the results which have endured the time control and were somewhat extended and supplemented in the light of the last achievements in the corresponding areas of science. And, finally, this book fills in, by quite a natural way, the peculiar gap in the structure of computational aerodynamics, connected with a statistical modeling.

The book was carried out within the frame of a scientific project "POISK" ("Search"), elaborated at the Department of Aeromechanics and Flying Technique of the Moscow Physico-Technical Institute (MPTI). The essence of this project consists in the following. All around the world the tremendous number of researches is working on the solution of fundamental and applied problems, connected with turbulence, especially with nonuniform and anisotropic one. Accumulated is the tremendous volume of factual material, and as rather actual point became that of preparation of a guide-book for orientation in that boundless sea of the theoretical, experimental, and numerical results. At the above-mentioned Department of MPTI was developed the project for such a guide-book and accompanying materials. The project's structure presents in itself the creation of books, containing the analysis of experimental results, of the theoretical and computer-based methods. This project is already partly realized.

In particular, published is the book surveying the contemporary experimental research on the dynamical structures within the turbulent boundary layer:

(1) Yu.I. Khlopkov, V.A. Zharov, S.L. Gorelov, Coherent Structures in the Turbulent Boundary Layer. M., MPTI, 2002.
Presented in this book, containing over 400 references, are the principles of physics of the dynamical processes in turbulent boundary layer, such as the phenomenon of bursting, the formation of streaks, and the processes of transfer of momentum and energy from the outer boundary of layer to that of the flow itself. Moreover, presented is the critical analysis of the foreign experimental works, formulated are the actual problems. As it was found, the analysis of experimental investigations, conducted during a prolonged period (over 40 years), revealed those essential features of the flows of fluid and gas, which might be used by the construction of a general theory of the processes involved. The theoretical studies of turbulent flows are carried out during a long time, too. The considerable part of that time was devoted to the search of the most effective methods of problem's solution. In the survey book,

(2) Yu.I. Khlopkov, V.A. Zharov, S.L. Gorelov, Lectures on the Theoretical Methods of Study of Turbulence. M., MPTI, 2005, are summed the results of these studies, and presented is the criticism of various methods, used at the earlier stage of the development of a theory. Thus, the reader is permitted to orientate himself in the contemporary directions of study.
The publishing house of MPTI has published also the survey book,

(3) Yu.I. Khlopkov, V.A. Zharov, S.L. Gorelov, Renormgroup Methods of the Description of Turbulent Motions of Incompressible Fluid. M., MPTI, 2006.
Presented in this book is the survey of results of elaboration and application of the number of methods, named as renormgroup methods, for the construction of models of turbulent flows of the incompressible fluid, both in the uniform and isotropic case, and in the case of a strong anisotropy and nonuniformity. The book is based on the studying of about 1000 of the original works, selected

from the totality of which were the most actual ones, according to the authors opinion. The largest part of contents is devoted to the three sub-network models of turbulence, which are widely used in the contemporary practical activity of various specialists in aerodynamics. The book is published as a textbook for students, though it demands the considerable efforts for its understanding and is intended, actually, for the professors and postgraduates. At the present moment is prepared for publication "The Lecture Course on the Theory of Turbulence", which was presented at the Department of Aeromechanics and Flying Technique by the Professor V.N. Zhigulev and is devoted to the studies on that problem on the kinetical level. Further on, it is planned to carry out the survey and analysis of modern numerical methods, used by the modeling of complicated unsteady flows of fluid and gas. The authors are expressing their deepest gratitude to the Russian Foundation of Fundamental Research, which is supporting this project, especially useful for the young generation.

The authors thank their colleagues M.N. Kogan, V.A. Zharov, S.L. Gorelov, I.V. Voronich, I.I. Lipatov, K.Yu. Gusarov, G.A. Tirskii, and V.V. Zasypalov for the participation in useful discussions and the observations spoken out, as well as the postgraduates Olga Rovenskaja, Andrei Bukin, Tatjana Stanko, Anton Khlopkov, Zei Yahr, Tun Tun, and Ignat Ikrjanov for their help in our work. Our thanks also to Marina Spirkina and Valentina Druzhinina for their help in putting the manuscript into shape.

Their special gratitude authors express to Dr Vsevolod Pavlovich Shidlovsky for his qualified labor on the translation of this book, and to the editors of translation — Natalja Nosova and Irina Tarkhanova.

Contents

0. Introduction

0.1 The General Scheme of Monte Carlo Methods

The first publication on the use of Monte Carlo methods was made by Hall[1] in 1873 while organizing the stochastic process of experimental determination of π number by means of throwing of needle on the sheet of lined paper. The bright example of the application of Monte Carlo methods consists in the use of idea by J. von Neumann realized in 1940s of the past century, by the modeling of the neutron's trajectories in Los Alamos laboratory. In spite of the fact that Monte Carlo methods are connected with a large amount of computations, the lack of electronical computational technique did not embarrass investigators in neither case during their application of the methods of statistical modeling, because in both cases, the study was concerned with realization of accidental processes. And these methods acquired their romantic title by the name of enclave of Monaco, which is famous due to its gambling houses, where the principal object is roulette — the most perfect instrument for obtaining the accidental numbers. And the first publication with a systematic exposition of that matter was made in 1949 by Metropolis and Ulam,[2] where Monte Carlo method was applied to a solution of linear integral equations. And in that publication was implicitly revealed the problem of the passing of neutrons through the substance. When speaking of Russia, the publications on Monte Carlo methods began to appear actively after the International Conference in Geneva devoted to the peaceful applications of nuclear energy. As one of the first one could mention the work by Vladimirov and Sobol.[3] Beginning with the earliest 1970s, side by side with the regular methods those of Monte Carlo obtained their proper place in computational mathematics (Marchuk, Samarskii, Popov, Belotserkovskii, Bakhvalov, Ermakov, Mikhailov, Sobol, Bird, Haviland, Kogan, Perepukhov, and Janitskii).

The general scheme of Monte Carlo method is based on the central limiting theorem of the theory of probability, which states that the accidental quantity $Y = \sum_{i=1}^{n} X_i$ which is equal to a sum of the large amount N of arbitrary accidental numbers X_i having identical mathematical expectations m and dispersions σ^2, is always distributed according to a normal law with mathematical expectation $N \cdot m$ and dispersion $N \cdot \sigma^2$.

Let us assume that we wish to find a solution of some equation or a result of some process I. If the accidental quantity ξ with a probability density p is built up in such a way that the mathematical expectation for this quantity would be equal to the solution we are looking for, $M(\xi) = I$, then that would give the simple way of estimation of both the solution and the error,

$$I = M(\xi) \approx \frac{1}{N} \sum_{i=1}^{N} \xi_i \pm \frac{3\sigma}{\sqrt{N}}.$$

Following from above are the general properties of the methods:

— the absolute convergence to a solution as $1/N$;
— the strong dependence of the error ε on the number of trials, as $\varepsilon \approx \frac{1}{\sqrt{N}}$ (that is, for the diminishment of the error on one order it is necessary to increase the number of trials on two orders);
— the main way of the error's diminishment consists in the maximal diminishment of dispersion or, in the other words, it is necessary to draw as near as possible the probability density $p(x)$ of the accidental quantity ξ to the mathematical formulation of the problem or to the physics of a phenomenon modeled;
— the error does not react on the problem's dimensionality (by the use of finite-difference methods the transition from the one-dimensional problem to the three-dimensional one the number of computations would be increased on two orders, while in Monte Carlo methods the number of computations remains on the same order);
— the simple structure of computational algorithm (number N of the single-type computations by the realization of accidental quantity);
— moreover, the construction of accidental quantity ξ might be based on the physical nature of process only, and would not demand, as it is in the regular methods, the compulsory formulation of the equation; such a quality becomes more and more actual for modern problems.

The main properties of Monte Carlo methods, as well as the conditions at which they yield or surpass the traditional finite-difference approaches, might be demonstrated by the application to some simple problem, for example, to the problem of the computation of an integral

$$I = \int_a^b f(x)dx,$$

where **x**, **a**, and **b** are vectors in a n-dimensional Euclidian space. Let us build up the accidental quantity ξ with density $p(x)$ in such a way that the mathematical expectation

$$M(\xi) = \int_{-\infty}^{\infty} \xi \cdot p(x)dx,$$

would occur to be equal to our integral I. Then, if within the proper limits one would choose $\xi = f(x)/p(x)$, then the central limiting theorem would give

$$I = \frac{1}{N} \sum_{i=1}^{N} \xi_i \pm \frac{3\varepsilon}{\sqrt{N}}.$$

Thus, we have as *the first*: The computation of the integral I might be interpreted, from one side, as a solution of mathematically formulated problem, and, from the other side, as a direct modeling of determination of a volume found under the function $f(x)$.

The second: The computation of the one-dimensional integral I_1 by the Monte Carlo method corresponds to the integral's computation by the method of rectangles with a step $\Delta x \approx 1/N$ and an error $O(\Delta x)$. In principle, by the sufficiently good function $f(x)$ in one-dimensional case the integral I_1 might be calculated with accuracy $O(\Delta x^2)$ with trapezoids, with accuracy $O(\Delta x^3)$ with parabola's, and, generally speaking, with any predicted accuracy. In multi-dimensional case, the difficulties of use of the high-order schemes become to acquire such an essential character that by the computation of n-dimensional integrals I_n with $n \geq 3$ the high-order schemes are used just very rarely.

Let us build up the correspondence in effectiveness between regular methods and statistical ones. Let it be that n is the problem's dimensionality, Y — the number of knots at the axis, $R = Y^n$ — the total number of knots for regular methods, q — the order of accuracy of the scheme, N — the number of statistical trials, v — the number of operations for a treatment of a single knot, $\varepsilon_L = Y^{-q}$ — the error of computations for regular methods, $\varepsilon_K = N^{-1/2}$ — the error of computations for Monte Carlo, $L(\varepsilon) = v \cdot R = v \cdot \varepsilon^{-n/q}$ — the number of operations by the problem's solution with regular methods, $K(\varepsilon) = v \cdot N = v \cdot \varepsilon^{-2}$ — the number of operations by the use of Monte Carlo method. For the case of one and the same number of operations by the computation of solution by one or another method one would obtain the relation $n = 2q$. This means that with $n \geq 3$, when mainly the first-order schemes are used, the Monte Carlo methods occur to be preferable.

0.2 Special Position of Monte Carlo Methods in Computational Aerodynamics

Dynamics of the rarefied gases is treated by means of well-known integro-differential equation — *Boltzmann* equation:

$$\frac{\partial f}{\partial t} + \vec{\xi}\nabla f = \int (f' \cdot f_1' - f \cdot f_1) \cdot \vec{g} \cdot b \cdot db \cdot d\varepsilon \cdot d\vec{\xi}_1, \qquad (0.2.1)$$

where $f = f(t, x, y, z, \xi_x, \xi_y, \xi_z)$ is the function of molecule's distribution in respect of time, coordinates, and velocities, f', f_1' — distribution functions corresponding to the molecule's velocities after collision, ξ', ξ_1', $\vec{g}$ — relative velocities of molecules by collisions in pairs, $\vec{g} = \vec{\xi} - \vec{\xi}_1 = \vec{\xi}' - \vec{\xi}_1'$, b, and ε — aiming distance and azimuthal angle by the collisions of particles. The complicated nonlinear structure of the integral of collisions and large number of variables (in the general case — 7) do create the essential difficulties for the analysis, including the numerical one, and, practically, lead to the exclusion of the finite-differential approach from the process of the solution of serious problems. At the same time, the multidimensionality and probabilistical nature of kinetic processes create the natural ground for the application of Monte Carlo methods.

Historically, the application of Monte Carlo methods to the computational aerodynamics was initiated in TSAGI by the pioneering works by M.N. Kogan and V.A. Perepukhov devoted to the modeling of free-molecular flows about space objects, in the part of trajectory of their orbital flight. Such a modeling is just the simplest form of rarefied gas dynamics. The further development of the statistical computational methods was realized in the following three directions:

Professor M.N. Kogan. 1965.

— use of the Monte Carlo methods for the calculation of collision integrals found in the regular finite-difference schemes designed for the solution of kinetic equations;

— direct statistical modeling of the physical phenomenon which is splitted in two approaches: modeling of the trajectories of "trial particles" according to Haviland[4] and modeling of the evolution of "ensemble of particles" according to Bird[5];

— construction of the accidental process of the type of a procedure by Ulam and Neumann described in Ref. 6 and corresponding to a solution either of linearized kinetic equation,[8] or of Master Equation by Kac.[7]

The probabilistical nature of the aerodynamics of rarefied gases, which is so important for application and development of the numerical schemes of Monte Carlo, follows quite naturally from the general principles of kinetic theory and statistical physics.

Laureate of State prize
V.A. Perepukhov. 1965

The reasoning cited below might be, quite perfectly, looked at as the levels of completeness of description of a large molecular system. Further on, these levels will be needed for a construction of the effective methods of statistical modeling. The most detailed level of description is presented by a dynamical system. To describe such a system which comprises of a large number of elements N (note that molecular gas is just such a system with $N \approx 10^{23}$ molecules), it is necessary to

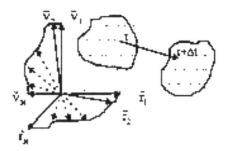

Fig. 0.1   Evolution of the dynamical system in $6N$-dimensional space.

set the initial coordinates and velocities of each molecule $(\vec{r}_j, \vec{v}_j)$ and the equations of evolution of this system:

$$m\frac{d^2\vec{r}_j}{dt^2} = \sum_{i\neq j}^{N} R_{ij}. \tag{0.2.2}$$

Solution of such a system appears to be quite unreal problem, even for a strongly rarefied gas — at the height of 400–600 km (the most popular orbits of satellites) one cubic centimeter contains 10^9 molecules. For this reason one comes to the less complete, that is, statistical description of the behavior of the system. In accordance to a Gibbs formalism one considers not a single system, but the ensemble of them in $6N$-dimensional Γ-space (Fig. 0.1), with system's distributed according to the N-particle distribution function $F(t, \vec{r}_1, \ldots, \vec{r}_N, \vec{v}_1, \ldots, \vec{v}_N) = F_N$, of which the sense is that of a probability for a system to be in the time moment t at the point $\vec{r}_1, \vec{r}_2, \ldots, \vec{r}_N, \vec{v}_1, \vec{v}_2, \ldots, \vec{v}_N$, in vicinity of $d\vec{r}_1, \ldots, d\vec{r}_N d\vec{v}, \ldots, d\vec{v}_N$ we have

$$dW = F_N d\vec{r}_1, \ldots, d\vec{r}_N d\vec{v}_1, \ldots, d\vec{v}_N.$$

Such an ensemble is described by the famous *Liouville* equation:

$$\frac{\partial F_N}{\partial t} + \sum_{i=1}^{N} v_i \frac{\partial F_N}{\partial r_i} + \sum_{i\neq j}^{N}\sum_{i=1}^{N} \frac{R_{ij}\partial F_N}{m\partial v_i} = 0. \tag{0.2.3}$$

And beginning with that moment the *Liouville* equation and all other kinetic equations following from the *Bogoljubov's* chain, including the last link — *Boltzmann* equation — possess the probabilistical nature. And in spite of the fact that Eq. (0.2.3) is simpler than system (0.2.2), it takes into consideration the N-particle collisions of molecules and also remains to be extremely complicated for a practical analysis. The transition to a less detailed level of description is connected with the further coarsening of system's description with the help of s-particle distribution functions $F_s = \int F_N d\vec{r}_{s+1} \cdots d\vec{r}_N d\vec{v}_{s+1} \cdots d\vec{v}_N$, which determine the probability of the simultaneous revelation of s particles independently of the state of the remaining $N - s$ particles. Following the ideas of *Bogoljubov* one obtains the chain of interconnected equations:

$$\frac{\partial F_s}{\partial t} + \sum_{i=1}^{s} v_i \frac{\partial F_s}{\partial r_i} + \sum_{0=1}^{s} \sum_{j \neq i}^{s} \frac{R_{ij} \partial F_s}{m \partial v_i} = - \sum_{i=1}^{s} (N-s) \frac{\partial}{\partial v_i} \int \frac{R_{i,s+1}}{m} F_{s+1} dr_{s+1} dv_{s+1},$$

(0.2.4)

up to the one-particle distribution function $F_1 = f(t, \vec{r}, \vec{\xi})$ for the Boltzmann's gas, taking into account only by-pair collisions:

$$\frac{\partial f}{\partial t} + \vec{\xi} \frac{\partial f}{\partial \vec{r}} + \frac{R_{12} \partial f}{m \partial \vec{\xi}} = - \frac{\partial}{\partial \vec{\xi}} \int \frac{R_{12}}{m} F_2 d\vec{r}_1 d\vec{\xi}_1.$$

Following Boltzmann we shall consider the molecules to be spherically symmetrical and, adopting the hypothesis of a molecular chaos, $F_2(t, \vec{r}, \vec{v}_1, \vec{v}) = F_1(t, \vec{r}, \vec{v}_1) F_1(t, \vec{r}, \vec{v}_2)$, one obtains Eq. (0.2.1).

As rather interesting one might consider the particular case of Liouville's equation (0.2.3) and Bogoljubov's chain (0.2.4) for the spatially uniform gas consisting of the limited number of particles. At the terminal link this case leads to obtaining the famous equation by *Kac* — "Master Equation"[7]:

$$\frac{\partial \phi_1(t, \vec{\xi}_1)}{\partial t} = \frac{N-1}{N} \int [\phi_2(t, \vec{\xi}_1', \vec{\xi}_2') = \phi_2(t, \vec{\xi}_1, \vec{\xi}_2)] \cdot g_{12} d\sigma_{12} d\vec{\xi}, \quad (0.2.5)$$

where ϕ_1 and ϕ_2 are one- and two-particle distribution functions. Unlike the Boltzmann's equation, Eq. (0.2.5) is linear, and this fact will be used by the construction and estimation of the effective computational schemes of direct statistical modeling. When coming back to the Boltzmann equation, from the determination of the function f it would be easy to obtain all the macroscopic parameters. Thus, the number n of molecules within the unit volume of gas is equal to

$$n(t, x) = \int f(t, x, \xi) d\xi.$$

Similarly to that, the mean velocity of molecules, stress tensor and vector of the flow of energy are defined by the relations

$$u(t, x) = (1/n) \int \xi f(t, x, \xi) d\xi,$$

$$P_{ij} = m \int c_i c_j f(t, x, \xi) d\xi,$$

$$q_i = (m/2) \int c^2 c_i f(t, x, \xi) d\xi,$$

where $c = \xi - u$ is thermal velocity of molecules. The mean energy of the thermal motion of molecules is usually characterized by temperature

$$\frac{3}{2}kT = \frac{1}{n} \int \frac{mc^2}{2} f(t, x, \xi)d\xi.$$

By way of application to the Boltzmann equation procedure of Enskog and Chapman one obtains the hydrodynamical level of description. Thus, on that level the description corresponds to Navier–Stokes equations:

$$\frac{\partial \rho}{\partial e} + \frac{\partial \rho u_i}{\partial x_i} = 0,$$

$$\left(\frac{\partial}{\partial t} + u_j \frac{\partial}{\partial x_j}\right) u_i = \frac{1}{\rho} \frac{\partial P_{ij}}{\partial x_j} + \frac{X}{m},$$

$$\frac{3}{2}R\rho \left(\frac{\partial}{\partial t} + u_j \frac{\partial}{\partial x_j}\right) T = -\frac{\partial q_j}{\partial x_j} - P_{ij}\frac{\partial u_j}{\partial x_j},$$

$$P_{ij} = p_{ij} + \rho_{ij}p, \quad p = \rho RT, \qquad (0.2.6)$$

$$p_{ij} = \mu \left(\frac{\partial u_i}{\partial x_j} + \frac{\partial u_j}{\partial x_i} - \frac{2}{3}\delta_{ij}\frac{\partial u_r}{\partial x_r}\right),$$

$$q_i = -\lambda \frac{\partial T}{\partial x_i}$$

and Euler's equations:

$$\frac{\partial \rho}{\partial t} + \frac{\partial \rho u_i}{\partial x_i} = 0,$$

$$\left(\frac{\partial}{\partial t} + u_j \frac{\partial}{\partial x_j}\right) u_i = -\frac{1}{\rho}\text{grad} \cdot p, \qquad (0.2.7)$$

$$\frac{3}{2}R\rho \left(\frac{\partial}{\partial t} + u_j \frac{\partial}{\partial x_j}\right) T = -p \cdot \text{div}u,$$

$$p = \rho RT.$$

Following the general logic of the present exposition one might assume that dynamics of continuum, as a particular case of kinetical approach to the treatment of motion of a gas, possesses the features of statistical nature and permits the realization of statistical modeling, just what will be demonstrated below.

0.3   The Position of Monte Carlo Methods in Modern Mathematics

The singular features of Monte Carlo methods cited in the preceding section just lead to the necessity of marking the position of methods of the statistical modeling in modern mathematics. Undoubtedly, at the present moment the priority is firmly kept by the traditional theoretical approaches (see works by Sadovnichij,[258,259] Matrosov,[270] Zhuravlev, Fljorov,[285] and others) and finite-differential approaches (see Marchuk,[236] Bakhvalov,[235,247] Kholodov, Magomedov,[249] Popov, Samarskii,[252,254] Rudakov,[264] Rusanov,[265,282] Kostomarov, Tikhonov,[260,262,271] Konovalov,[274] and others) to the solution of equations of mathematical physics. But permanently in larger and larger degree are beginning to come true the prophetical words of academician Vladimirov spoken out by him more than half a century ago: "... will come the time when along the development of the computational technique the Monte Carlo methods will the more and more successfully compete with the traditional ones". And this time has already come in reality. The Monte Carlo methods have safely obtained their place in the list of traditional methods of theoretical and applied mathematics (see Vladimirov,[3] Marchuk, Ermakov, Mikhailov, Sobol,[6,33–40,236,237] Bird, Haviland,[4,5] Kogan, Perepukhov,[8,51] Belotserkovskii, Yanitskii[9]). It is likely that this tendency of expansion of Monte Carlo methods into the realm of modern mathematics will steadily last. Just at the present time there exists the whole number of regimes of the gaseous flows, which cannot be properly analyzed with the help of network algorithms. That is the class of problems connected with the simulation of flows in high-altitude hypersonic aerodynamics and with solution of the Boltzmann equation. Moreover, one can indicate certain cases, as, for example, the flow of rarefied gas about thin cold bodies, where Monte Carlo methods prove to be the only source of information on the aerodynamical characteristics of such bodies. It is impossible to simulate flows of such a type in the wind tunnels presently existing.

The modern stage of development of the computational technique, of the computational methods and of the programming is, first of all, connected with perfection of the parallel computations (see Evtushenko,[256] Koroljov,[271,273] Voevodin, Voevodin,[24] Ivannikov,[239,248] Zabrodin,[239] Pavlovskii[268]). The powers of modern computers reach the figures of hundreds of teraflops. The most powerful in the world is the platform BLUEGENE/L, created by the IBM company and projected for 596 teraflop. Accepted for realization in USA is the 5-year project uniting the efforts of the companies SANDIA, OAK RIDGE, Ministry of Energetics and the number of other ministries, aimed to a creation of the computational complex with the power of one exaflop (one million of teraflops). In this connection, the singular features of Monte Carlo methods indicated above acquire quite a new tincture. For this reason the present book cannot avoid the consideration of the prospect of parallelization of the algorithms of statistical modeling.

Nonuniform and anisotropical turbulent flows of fluid and gas represent the most frequently met in nature forms of the motion of matter. Such flows are met in micro-, macro- and megaworld. The laws of turbulent motions call forth the processes in nanotechnologies, the dynamics of flight of flying apparatuses, influence the geophysical processes, the climatic changes, the evolution of stellar accumulations (see works by Velikhov, Popov, Belotserkovskii, Berdyshev, Vasin, Dymnikov, Iljin, Kholodov, Gushchin, Zhuravlev, Zhizhchenko, Chechetkin[241–246,249–253,278,279,283]). The hope of using for the description of complicated unsteady flows the permanently increasing powers of the modern computers was not justified — just as it was warned 40 years ago by Anatilii Alexeevich Dorodnicyn. The absence of the adequate physical models for description of complicated flows of the fluid and gas cannot be substituted by teraflops of the modern processors. Therefore, in this book much attention is given to the methods of formation of the physical models of turbulence. Extremely useful proves to be the experience of formation of the models based on the general principles in various areas of physics, climatology, geophysics, navigation, turbulence (works by Velikhov, Popov, Dymnikov, Zhizhchenko, Petrov, Krasnoshchekov, Kostomarov, Pavlovskii, Savin, Vasin, Berdyshev, Malinetskii, Alexeev[234,252–255,261,268,275–279,283,286]). Presented in this book are the original models of turbulence, such as kinetical molecular and fluidical models and the model of three-wave resonance.

The volume of this book does not permit to make a comparison of the methods of statistical modeling with such an important section of mathematical science as optimization problems which are closely connected with a solution of large systems of the algebraic equations (see Zhuravlev, Iljin, Moiseev, Evtushenko, Matrosov, Arutjunov, Eremin, Tyrtyshnikov[245,246,250,251,257,263,266–269,287]). These questions were considered in the preceding books of the present authors (see Khlopkov, Gorelov[288,289]), which, unfortunately, became at the present moment the bibliographical rarity. Therefore, it is planned to come back to the questions of solution of the linear and nonlinear problems of algebra and of mathematical physics, to the optimizational and extremal problems, and it is planned to devote to these areas of science some special editions.

0.4 Short Survey of Monte Carlo Methods in Computational Aerodynamics

The great scientifical and applied importance possessed presently by the dynamics of rarefied gases (DRG) is explained by the practical importance of the vast area of problems connected with the modern stage of mastering the space, with the development of vacuum technology, of laser technique, and with the other branches of scientifical and technical progress. The methods developed in rarefied gas dynamics are widely applied to the solution of problems not connected with the rarefaction of matter — the theory of homogeneous and heterogeneous processes, the theory of evaporational and adsorptional processes, of the nonequilibrium flows, of the prescription and setting of the boundary conditions and coefficients of transfer in mechanics of continuum. The necessity of qualitative and numerical analysis of the phenomena in rarefied gas dynamics, the complexity, and multi-dimensionality of the equations which are to be dealt with, stimulated the development of effective and original numerical methods. The special features of physical phenomena which are met in DRG and of equations describing these phenomena (those are, mainly, equations of Boltzmann and of Navier–Stokes) lead to the imposition of the multitude of demands on the methods under development:

— Justification of the computational procedure and accuracy of the solution resulted, aimed at the use of that procedure as model one.
— The possibility of the effective modeling of complicated flows, such as three-dimensional flow about bodies in all the ranges of transitional regime, beginning with the free-molecular one and finishing with one for continuum, as well as taking into account physical and chemical properties of gases.

In accordance with the main demands the methods existing might be united into groups, determined by the degrees of their justification and by their connection with kinetical equations, as well as by degree of possibility of modeling the complicated phenomena. For inclusion into *the first group* one might mention, first of all, the regular methods — classical finite-difference approaches; semiregular methods with the use of Monte Carlo procedures for computation of collision integrals; and, finally, statistical procedures of the type of Ulam–Neumann, aimed at the solution of the kinetic equations. For inclusion into *the second group* might be singled out the methods of direct statistical modeling of the real flows. It is necessary to note that such a subdivision of methods occurs to be rather conditional, and between many of these methods is established an interconnection. According to our opinion, the main point in classification is, after all, the effectivity of solution of the complicated problems.

The concise retrospective of the development of methods might be presented in the following way. In the monograph by *Kogan*[8] are presented the main computational methods of rarefied gas dynamics developed up to the mid-1960s. The analysis of methods developed somewhat later was given in the survey paper by Belotserkovskii,[9] while the detailed description of the main approaches of statistical modeling is presented in the collection of papers and in monographs by Bird and Belotserkovskii.[10,11,12] The description of regular and semiregular methods was given by *Ryzhov*.[13] As it was already noted, speaking of the degree of justification of computational procedure one should mention, first of all, the regular

methods. Generally speaking, the essence of these methods consists in approximation of the distribution function by its values in the points of phase space and in
subsequent solution of the difference equations. In addition to the natural demand
of the ability of conservation of the large volume of information and of operation
with that volume, the serious difficulties arise in connection with finite-differential
approximation of the collision integrals. Just for that reason the natural stage of
the development of methods was in many cases reduced to the use of Monte Carlo
procedures for the computation of collision integrals. By application of these methods to Boltzmann equation a number of model solutions was obtained, mainly
for spatially uniform and one-dimensional flows, as for example, by *Ender M.,
Ender A., Tcheremissin,*[14,15] and for two-dimensional problems by *Tcheremissin,
Shcherbak.*[6,17] By the solution of modified kinetic equations the possibilities of
methods are essentially amplified — see *Krook, Holway, Shakhov, Larina, Rykov,
Limar, Huang.*[18–26] The substitution of the collision integral by some simplified
expression does in many cases permit to build up the computational procedure for
collision integral at the level of macroparameters, and in such a way the effectiveness of the methods applied is considerably increased — see *Krook, Holway,
Shakhov.*[18–20]

J. Bird and O. Belotserkovskii. XIII International
Symposium on dynamics of rarefied gas. Novosibirsk, 1983.

The complicated multi-dimensional structure of kinetic equations, on one hand,
and the abundance of information brought by the distribution function, on the other
hand, stimulated the application and development of statistical procedures for the
solution of problems in rarefied gas dynamics. The first application of statistical
methods was connected with direct modeling of gaseous flows — see *Haviland,
Bird.*[27–32] It should be immediately noted that the methods of direct statistical
modeling proved to be the most effective in rarefied gas dynamics. In addition to
reasons mentioned earlier, this property might be explained by the statistical nature

of kinetic equations. Among the methods of direct statistical modeling one is able to single out two approaches: the method of stationary direct statistical modeling by *Haviland*[27−29] and the method of nonstationary direct statistical modeling by *Bird*.[30−32] Since this work is devoted, mainly, to the statistical methods, let us briefly describe the essence of approaches mentioned above.

By the use of either of this methods, within the area of flow is chosen a certain volume, on the boundaries of which is by the usual way prescribed the form of distribution function and the law of interaction of gas with a surface. The area of computation is divided in cells, within each of which the distribution function is modeled by some quantity of particles. The sizes of cells are chosen on the basis of condition of constancy of distribution function over the cell's volume. The evolution of particles is divided in time in small intervals of the length Δt chosen from a condition $\vec{\xi}\Delta t \ll \vec{\lambda}$, where $\vec{\xi}$ and $\vec{\lambda}$ are characteristical values of velocity and mean free path of molecules. The main distinctive feature of the procedures under consideration consists in the fact that in the case of nonstationary modeling is realized the simultaneous watching at the whole ensemble of particles. This behavior permits to construct the particle's trajectories taking into account the varying in time frequency of collisions and leads to some stabilization process. In the case of stationary modeling, the watching is realized for singular, so called trial, particles, which in fact leads to a necessity of knowing the distribution function for field particles, and, correspondingly, to some iterational process. Both approaches admit the number of modifications of the principal character, which are essentially increasing their effectiveness and lead to their successful use for three-dimensional problems. Generally speaking, the direct modeling of gaseous flows proves to be the universal tool of studies not only in the area of rarefied gas, but also, as it is shown in the present book, in mechanics of continuum. However, one of the positive qualities of direct modeling — solution of the problems without addressing to the equations — proves sometimes to be also its main negative property. The lack of a direct connection with the equation describing the process evokes the certain mistrust to the results obtained and leads to certain difficulties in systematic approach to the increase of method's effectiveness. For this reason, the works on justification and settlement of correlation between statistical procedures controlling equation. The well-known statistical procedure by Ulam — Neumann for the solution of integral equations and widely applied in the theory of radiation (see *Marchuk, Ermakov, Mikhailov, Sobol*[33−35]) might be directly applied to kinetic equations only in the case of linearized kinetical equation (see *Khlopkov*[36]). In nonlinear case (see *Ermakov, Nephedov*[37,38]) the modification of procedure by Ulam–Neumann was proposed on the basis of theory of branching processes. However, the practical realization of this method was connected with large amount of computations and proved to be rather difficult. The construction of standard procedure by Ulam and Neumann for nonlinear equation demands to carry out the artificial linearization, and this leads to an iterative process, to keep in memory the information on preceding iteration, and, correspondingly, to a division of phase space in cells, which leads to appearance of an additional source of errors and does not enter as necessary part into methods by *Khlopkov, Ivanov* and *Grigorjev*.[39,40] The procedure, constructed in such a way, corresponds to the method of stationary statistical modeling and this fact was used for the formation and justification of the methods by *Khlopkov* and *Ivanov*.[39,40] A somewhat more complicated situation arises in connection with justification of the method of nonstationary modeling. Devoted to the determination of its connection with a solution of kinetic equations are the works by *Belotserkovskii* and *Yanitskii*.[9,12] After all, in the practical realization the preference is rendered to the methods of direct statistical modeling. Just with the help of these methods were solved the most of extremely complicated problems of practical importance. The modernization of the methods of stationary statistical modeling was realized, mainly, by way of cutting down the operative memory of a computer. Thus, in the papers by *Vlasov*[41,42] proposed is the procedure of construction of trajectories which does not demand conservation in memory of the distribution function. The corresponding idea is based on the fact that density of probability of the velocity of a field molecule is equal to the distribution function we are looking for, which is normalized over unity. As the other direction of increase of the method's effectiveness appears the approximation of the distribution function of field particles with the help of a certain number of moments. The vast possibilities for the improvement of the methods of stationary modeling are found by means of use of the model kinetic equations (see *Khlopkov, Ivanov*[39,40]). In this case, the realization of the procedure of computation of collisions does not demand a knowledge of the distribution function of the field particles, because after collision the trial particle acquires the velocity corresponding to the distribution function in equilibrium.

As the central point in the method of nonstationary statistical modeling appears the procedure of calculation of collisions. The pair of particles is chosen for collision in the correspondence to the frequency of molecular collisions, but independently of the distance between particles within the cell considered. The velocities of particles after collision are chosen in correspondence with the laws of molecular interaction. In spite of the fact that method's effectiveness depends on comparatively numerous amount of the parameters of calculation scheme (stabilization, splitting in time, achievement of the steady regime, step in time, step of the spatial net, and so on), the main works on the method's improvement are devoted to the perfection of a collisional procedure and to the diminishment of a statistical error of the scheme as a principal factor permitting to diminish a number of particles in cells, and thus to diminish the operative memory of a computer. For example, in the paper by *Eropheev* and *Perepukhov*[43] was proposed a modification of collisional procedure for Maxwellian molecules, by the use of which the computational results are, practically, independent of the number of particles in cell when this number varies between 40 and 6 (in the traditional computations the number of particles in cell is of the order of 30). In the papers by *Belotserkovskii* and *Yanitskii*[44-49] proposed was the method, in which at the stage of collisions the subsystem of model particles is considered as the N-particle model of Kac. The modeling of collision is reduced to a statistical realization of the evolution of Kac's model during the period of time Δt. The time of collision in Kac's model is calculated in correspondence with statistics of collision in perfect gas. This scheme permits to use essentially smaller number of particles in cell and smaller step of computational network. The analysis of results of the calculations has shown that these results are practically independent on the number of particles in cell up to the value of 2. As it was already noted, in the practical realization for the problems of rarefied gas dynamics the statistical methods proved to be more effective in comparison with regular methods and semiregular ones. For the problems of flow about objects, which are the most essential in aerodynamics, these methods were, first of all, successfully applied by *Perepukhov*[51,52] for the obtainment of aerodynamical characteristics of various, including complicated, bodies in the free-molecular flow, and in the flow, close to free-molecular one. In this direction, with the help of statistical methods based on stationary and nonstationary approaches, one was able by the solution of problems of flow about objects to advance to the *Knudsen* numbers up to the order of 0.001 in the planar and axisymmetrical cases (see *Khlopkov, Eropheev, Perepukhov, Ivanov, Vlasov, Gorelov*[39,40,53-69]), as well as in three-dimensional cases (see *Eropheev, Kravchuk, Serov, Khlopkov, Ivanov*[70-76,100]). According to all the evidences, the application of statistical modeling in its traditional form with the aim of penetration in the area of the mechanics of continuum proves to be ineffective with the contemporary computational background. And since the contemporary stage of development of the aerocosmic technique demands, nevertheless, to obtain the information within the whole range of transitional regime, then the flows of slightly rarefied gases demand, for their treatment, that some special methods of solution would be developed. Thus, of a certain interest are the methods which use the information on distribution function obtained on the basis of the physical considerations connected with continuum character of the matter. One of such directions is reduced to a solution of the equations for moments, of which the order is higher than that of Euler and Navier–Stokes equations. Up to the present moment, however, stays to be open the question whether the obtained improvement of accuracy would overpower the considerable complication of the corresponding macroscopic equations and difficulties in setting the boundary conditions.

Participants of the International Symposium on dynamics of rarefied gas in Novosibirsk, 1983. At center — V. Yaniyskiy, Yu. Khlopkov, M. Ivanov.

The other direction of investigation is connected with a simultaneous solution at the subsequently posed elementary intervals of time of both kinetic and continuum equations, thus obtaining the results leading to mutual complementations and improvements of accuracy, However, following in this direction one sees that to the difficulties of solution of the kinetic equations are, certainly, added the difficulties of solution of the Navier–Stokes equations, and just for that reason this method is applied only to one-dimensional problems. It seems to be more effective to single out within the bulk of a flow the certain areas with different physical properties, governed by the equations of different types, which are knitted together at the boundaries of these areas. Such a trick is widely used in mechanics of continuum (see *Sychev*[50]), but presently was not yet extensively spread in computational dynamics of rarefied gas and is presented only by some isolated works, like that by *Khlopkov*.[70] Furthermore, it seems to be extremely interesting and deserving of the further development that the computational methodologies developed in rarefied gas dynamics would be extended to such nontraditional areas of application as viscous and nonviscous flows of continuous matter. In particular, such a way of action permits to develop a unique computational procedure which would be independent of the degree of rarefaction of the matter.

In this connection of certain interest are the methods based on modeling the continuous matter by the ensemble of particles (see *Belotserkovskii, Harlow, Khlopkov,*

Kravchuk[12,70,71,74,77,79–82]), which would possess the proper peculiarities and the aggregate of which would characterize the matter under consideration. Thus, in methods by *Harlow, Gentry* and *Belotserkovskii*[12,79,80] the particles are representing the Lagrangian form of description of the flow of perfect gas, while in the papers by *Pullin, Khloplov* and *Kravchuk*[74,81] the continuum is modeled by the molecular distribution function. Generally speaking, the use of statistical approach in the description of a continuous matter with the help of the particles ensemble permits, on the one hand, to use the experience obtained in computational dynamics of rarefied gas, and, on the other hand, to develop a unified procedure of computation for all the regimes of flow (see *Kogan, Khlopkov*[82,83]). Finally, it is impossible not to note the practically important direction of the development of computational methods based on the hypothesis of locality (see *Barantsev, Galkin, Eropheev, Tolstykh, Bunimovich, Bass, Khlopkov*[84–96]). These methods appear as approximate and semi-empirical ones, permitting the effective obtainment of the integral aerodynamical characteristics of the apparatuses at all the regimes of flow.

J. Bird and Yu. Khlopkov. XXV International Symposium on dynamics of rarefied gas. Saint Petersburg, 2007.

0.5 Construction of the Effective Methods of Statistical Modeling

It is clear that at the present time the central place in studies on rarefied gas dynamics belongs, quite certainly, to the methods of direct statistical modeling. The main portion of each of these methods consists in modeling such a procedure of molecular collisions which would be able to increase just in several orders the computational speed and to diminish the volume of operative memory of the computer, when the comparison is made with its initial version. However, by the consideration and substantiation of the application of these methods it is impossible, practically, to avoid the survey of that kinetic equation which describes the phenomenon to be modeled. The formalization of connection between statistical procedure and solution of the kinetic equation is inevitable due to a number of reasons. First, it is needed for a solution to be trusted, and for the results obtained to be used as a sample ones, since the solutions of numerous typical problems were obtained initially just by the methods of direct modeling, and up to the present moment were not repeated with the help of other methods. Second, the determination of the interrelation between modeling and solution of the equation leads to a possibility of using the well-developed apparatus of the numerical, both regular and statistical, methods of solution of the equations of mathematical physics for the analysis of the methods and improvement of their effectiveness. And, third, such a way of action permits to formulate a certain general approach to the construction of methods and firmly excludes any false modifications of these.

It is to be stressed that on the way of development of the effective numerical algorithms the complexity of the practical problems of high-altitude aerodynamics demands, inevitably, to bring home the whole arsenal of the analytical, experimental, and numerical means of investigation of the flows of rarefied gases. In this connection, acquires the special value the analysis of kinetic equation and study of various models of it. Frequently used are the approximate representations of the collisional integral and of the distribution function. The most widespread approximate forms of kinetic equations are:

— The model equation by *Krook*[18]:

$$\frac{df}{dt} = \nu(f_0 - f), \tag{0.5.1}$$

where ν is the frequency of collisions,
$f_0 = n\left(\frac{m}{2\pi kT}\right)^{3/2} e^{-m/(2kT)(\vec{\xi}-\vec{v})^2}$ is the equilibrium distribution function.
— The ellipsoidal model by *Holway*[19]:

$$\frac{df}{dt} = \nu(f_e - f), \tag{0.5.2}$$

f_e — the ellipsoidal distribution function.
— The approximational model by *Shakhov*,[20] which, unlike the preceding models, gives the correct Prandtl number *Pr*:

$$\frac{df}{dt} = \nu(f^+ - f), \tag{0.5.3}$$

$f^+ = f_0\left[1 + \frac{4}{5}(1 - Pr)s_\alpha c_\alpha\left(c^2 - \frac{5}{2}\right)\right]$, $s_i = \frac{1}{n}\int c_i c^2 f d\vec{\xi}$, c is dimensionless molecular thermal velocity.

Just at the same point it is necessary to mention the linearized Boltzmann equation, which is rigorously deduced from the complete Boltzmann equation under the condition that distribution function is only slightly different from equilibrium one — see *Kogan*[8]:

$$\frac{d\phi}{dt} = k(\vec{\xi})\phi + \int L(\vec{\xi}, \vec{\xi}_1)\phi_1 d\vec{\xi}_1, \qquad (0.5.4)$$

where $f = f_0(1 + \varphi)$, $\varphi \ll 1$; $k(\vec{\xi})$ and $L(\vec{\xi}, \vec{\xi}_1)$ are some known functions of molecular velocities, dependent on the kind of particles.

In distinction of the linearized equation the model ones do not rigorously follow from the Boltzmann equation. Moreover, the model equations are found to be significantly more nonlinear than initial one, but in practical realization by the numerical modeling these equations might prove to be simpler. Naturally, by their practical realization the methods of direct statistical modeling based on approaches by Bird and Haviland proved to be more effective, and their modifications with alternating success realized their victorious train in computational aerodynamics. At the present moment, the unconditional priority belongs to the Bird's method. Due to the works of Russian scientists *O.Belotserkovskii. V. Yanitskii, M. Ivanov, V. Perepukhov* and *A. Eropheev*,[12,10–20] the modifications of Bird's method permitted to increase the effectiveness of that method, actually in several orders of value. The main feature of the method consists in the assumption that system's evolution at the small interval of time Δt is splitted in two quite obvious physical processes:

(1) relaxation in accordance to the collisional operator in kinetic equation:

$$\frac{\partial f}{\partial t} = J(f),$$

(2) free-molecular transfer,
(3)

$$\frac{\partial f}{\partial t} = -\vec{\xi}\nabla f.$$

This is well-known scheme of the first-order splitting in Δt for any arbitrary operator equation, but in the present case the advantage of scheme consists in the fact that it is splitting the dynamics of extremely complicated kinetical system in two obvious physical processes. The distribution function is modeled with the help of N particles, which at the first stage collide between themselves within each cell and in accordance with collisional frequency, during the time interval Δt, while at the second stage they fly during Δt, covering the distance $\vec{\xi}_o\Delta t$.

The central point in the method of nonstationary statistical modeling lies in procedure of calculation of the collisions. The pair of particles for collision is selected in accordance with frequency of molecular collisions and independently of the distance between these molecules within the cell chosen. The velocities of particles after the collision are selected in accordance with the laws of interaction between molecules. In spite of the fact that method's effectiveness depends on comparatively large number of parameters of the computational scheme (steadi-fication, splitting in time, taking out to the steady regime, time-step, type of the spatial network, and so on) the main body of work on the method's perfection was devoted to the improvement of collisional procedure and to the diminish-ment of a statistical error of the scheme as a main factor, which would permit to diminish the number of particles within the cells and, correspondingly, to dimin-ish the operative memory of a computer and to minimize the computational time. Thus, proposed in Ref. 18 was the modification of collisions for one particular case — Maxwellian molecules. By that modification the computational results do not, practically, depend on the number of particles within the cell, if that number varies between 40 and 6 (by the computations with the help of ordinary meth-ods this number is of the order of 30). Proposed in Refs. 11–16 is the general method, independent of the kind of molecules, in which at the stage of collisions the subsystem of particles within each cell is considered as the N-particle model by Kac[6]:

$$\frac{\partial \phi_1(t, \vec{\xi}_1)}{\partial t} = \frac{N-1}{N} \int \left[\phi_2(t, \vec{\xi}_1', \vec{\xi}_2') - \phi_2(t, \vec{\xi}_1, \vec{\xi}_2) \right] \cdot g_{12} d\sigma_{12} d\vec{\xi}_2.$$

Modeling of a collision is reduced to the statistical realization of evolution of the equation, however not of the Boltzmann's one (0.2.1), but of the Kac's model (0.2.5), during the interval Δt. The duration of collision in Kac's model is calculated in accordance with statistics of collisions in ideal gas consistent with the scheme of Bernoulli. This scheme permits to use essentially diminished number of particles within the cell and smaller size of step in the computational network. As it was shown by the analysis of results of the calculation, these results do not, practically, depend on the number of particles within the cell, up to the number of 2. The reason of that follows from the fact that Boltzmann equation demands, imperatively, the assumption of the molecular chaos to be fulfilled. However, with such a number of particles in cell, as is accessible in nowadays computers, the assumption of chaos is fulfilled only with a systematical error, whereas Eq. (0.2.5) is free of such a demand and, therefore, the stage of collision is computed as a purely Markovian process. From the other side, however, by $N \to \infty$ exists the complete equivalence between Kac's model and spatially uniform equation by Boltzmann.

Thus it might be said that the approach developed by Belotserkovskii and Yanitskii:

(1) Opens the way to a construction of the effective computational schemes, which would provide the possibility of solution of the three-dimensional problems of aerodynamical flow about objects.
(2) Permits to solve the most important methodological problem of the equivalence of a numerical method to a solution of the kinetic equation.

There exists a tremendous number of works on the traditional use of statistical modeling and, therefore, we shall limit ourselves, mainly, by the problems of aerodynamics. As it was already noted, in a practical realization of solution of the problems in rarefied gas dynamics the statistical methods proved to be more effective than regular ones and semiregular ones. For the problems of a flow about objects, which are the most essential in aerodynamics, these methods were, for the first time, successively applied to the obtainment of aerodynamical characteristics of various bodies, including complicated ones, in a free-molecular or nearly free-molecular flow. The corresponding methodics, developed more than 20 years ago, is presently brought to the state of standard programs and is widely used in the pertinent project and design institutions. The advance to lesser numbers Kn is connected with a sharp increase of computational difficulties due to a lesser free path of molecules and, consequently, to lesser steps in time and in space, and, in the case of direct modeling, due to an increased number of the particles modeling the distribution function.

As it was noted previously, at the present moment the indisputable priority belongs to the modern methods based on the Bird's approach to the modeling of dynamics of "molecular ensemble". But such a situation was not invariable. There was a period when the priority belonged to the method of "trial particles" based on the approach by Haviland. Probably, in the future will appear the effective scheme by this author, as it was already earlier, which would overpower those effective schemes by Bird's approach, which were mentioned above. Moreover, might appear, also, some new class of problems for which these schemes will prove to be preferable. But just at the present time one might single out some problems of physical chemistry for which the preference is given up to the method of trial particles. Certainly, there exists also a class of linear problems which happened to be, actually, the origin of this method.[2] And there is one obvious additional advantage of this method, consisting in the fact that, in distinction to the Bird's method, for it is not very important to obtain a complete identity of its solution and that of a kinetic equation. As concerns the computational aerodynamics, this method is based on the linearized Boltzmann's equation (0.5.4).

The conversion of computations into action on parallel lines with the help of high-performance supercomputer systems stays as one of the main directions in development of the modern computational mathematics (see *Voevodin, Voevodin, Ivannikov, Zabrodin, Kraiko*[226,231,238,239,240,248]). The supercomputer systems are permanently more and more widely used for a solution of fundamental and applied problems in the fields of nuclear physics, climatology, economics, pharmacology, modeling of the training devices and of virtual reality, computational aerodynamics. Due to the special qualities of Monte Carlo methods, in all the areas cited above the statistical modeling begins to play the ever-growing and outstanding part.

Chapter 1

The Main Equations and Approaches to Solutions of the Problems in Rarefied Gas Dynamics

1.1 The Main Equations in Rarefied Gas Dynamics

Dynamics of the rarefied gases is described by the well-known integro-differential kinetic equation of Boltzmann:

$$\frac{\partial f}{\partial t} + \xi \frac{df}{dx} = \int (f'f_1' - ff_1) g b \, db \, d\varepsilon \, d\bar{\xi}_1, \tag{1.1.1}$$

where f is function of distribution of molecules over velocities $\bar{\xi}$, $\bar{g}$ — relative velocities of molecules by binary collisions, b, ε — aiming distance and azimuthal angle by the collision of particles.

The complicated nonlinear structure of the integral of collisions and large number of variables (in the general case — 7) lead to the essential difficulties of the analysis, including numerical one, of the above equation, and, therefore, frequently used are the approximate presentations both of the collisional integral and of distribution function. The most widely used approximate forms of kinetic equations are:

— Model equation of Krook[18]:

$$\frac{df}{dt} = \upsilon(f_0 - f), \quad f_0 = n \left(\frac{m}{2\pi kT} \right)^{3/2} e^{-\frac{m}{2kT} (\bar{\xi} - \bar{\upsilon})^2}, \tag{1.1.2}$$

where υ is collisional frequency, f_0 — equilibrium distribution function.
— Ellipsoidal model of Holway[19]:

$$\frac{df}{dt} = \upsilon(f_e - f), \tag{1.1.3}$$

where f_e is ellipsoidal distribution function.
— Approximational model of Shakhov[20]:

$$\frac{df}{dt} = \upsilon(f^+ - f), \quad f^+ = f_0 \left[1 + \frac{4}{5}(1 - Pr)s_\alpha c_\alpha \left(c^2 - \frac{5}{2} \right) \right],$$

$$s_i = \frac{1}{n} \int c_i c^2 f \, d\bar{\xi}, \tag{1.1.4}$$

23

where c and Pr are nondimensional molecular thermal velocity and Prandtl number, correspondingly.

Just here one should necessarily mention the linearized Boltzmann's equation, which strictly follows from the complete Boltzmann's equation under the condition that distribution function is feebly different from the equilibrium one[8]:

$$\frac{d\varphi}{dt} = k(\bar{\xi})\varphi + \int K(\bar{\xi}, \bar{\xi}_1)\varphi_1 d\bar{\xi}_1, \tag{1.1.5}$$

where $f = f_0(1 + \varphi)$, $\varphi \ll 1$, $K(\bar{\xi})$, and $K(\bar{\xi}, \bar{\xi}_1)$ are some known functions of the molecular velocities, dependent on the particle's type.

In distinction of (1.1.5), Eqs. (1.1.2)–(1.1.4) cannot be rigorously obtained from the Boltzmann's equation and are, moreover, considerably more nonlinear than initial one. However, we shall see later, that in their practical realization by the numerical modeling these equations might prove to be simpler.

Considered in detail in Ref. 66 is the general methodics of the construction of kinetic models and their part, played in the investigation of the flows of rarefied gas. The basic scheme of the method, which might be properly considered as a method of approximation of Boltzmann's equation (1.1.1), consists in the procedure of substitution of the integral of collisions J by a certain simpler operator Q, which would realize only some limitation of iterational properties of Boltzmann's operation. Particularly, putted forward in the demand that for any prescribed function the first distributed moments of the approximating collisional operator and of the genuine one are, up to the certain order, coinciding. It is assumed that the approximate operator depends on the distribution functions, on the molecular velocity, as well as on the totality of macroparameters of the proper order. By this way appears the system of equations for determination of these macroparameters

$$a^{(0)}, a^{(1)}, \ldots, a^{(n)} :$$

$$\int J(f)\Psi d\bar{\xi} = \int Q(f, \bar{\xi}, a^{(0)}, a^{(1)}, \ldots, a^{(n)})\Psi d\bar{\xi}, \tag{1.1.6}$$

where $\Psi = 1, \xi_i, \xi_i\xi_i, \xi_1^{\alpha_1}, \xi_2^{\alpha_2}, \xi_3^{\alpha_3}, \ldots, \alpha_1 + \alpha_1 + \alpha_1 = m, m = 0, 1, 2, \ldots, n$.

The important property of nth approximation maintains that the obtained equations of moments up to the nth order are exactly coinciding with similar equations obtained from the Boltzmann's equation. The integrals in the relation (1.1.6) just represent the expansion of the usual demands on the fulfillment of conservation laws.

The approximate operator of collisions is presented in relaxational form

$$Q = \nu(f^+ - f), \tag{1.1.7}$$

while for the presentation of f^+ is used the locally Maxwellian expression multiplied by the proper polynomial:

$$f^+ = f_0\left(a^{(0)} + a_\alpha^{(1)}c_\alpha + a_{\alpha\beta}^{(2)}c_\alpha c_\beta + \cdots\right). \qquad (1.1.8)$$

Thus, by the proper limitation of the number of moments in (1.1.8) it would be possible to obtain, consequently, an operator (1.1.7) in the form of (1.1.2), (1.1.3), and (1.1.4), correspondingly, or to construct the model equation of a higher approximation.

It would be natural that any arbitrary simplification of the initial equation leads to some or another deviations from the genuine solution, which for various areas might be either essential or unessential. For example, Eq. (1.1.2) possesses the main properties of Boltzmann's equation: in particular, from that equation follows both conservation law and H-theorem, it is exactly transforming into Boltzmann's equation both in the free-molecular limit and in continuous medium limit without viscosity. However, by the transition to continuous medium the Krook's model gives an incorrect value of the Prandtl number. The models (1.1.3) and (1.1.4) are free from that defect, but at the same time the constructed model forms of kinetic equations are still sufficiently complicated and, practically, inaccessible for analytical treatment. For this reason, the main criterion in the process of the choice of some or another model by the study of typical problems is the comparison of numerical results, and just to that topic is devoted, mainly, the work by Shakhov and Khlopkov.[66]

1.2 The Main Approaches to the Construction of Statistical Algorithms

According to the usual consideration, the development of numerical statistical method is rarefied gas dynamics proceeds in the following three directions:

— Use of the Monte Carlo methods for calculation of the collisional integrals in regular finite-difference schemes of solution of the kinetic equations.
— Construction of the accidental process (of the type of procedure of Uhlam and Neumann), corresponding to the solution of kinetic equation.
— Direct statistical modeling of the physical phenomenon, which is divided in two approaches: stationary direct modeling and nonstationary one.

As it was already noted in the introduction, the central place in the dynamics of rarefied gases belongs to the methods of direct statistical modeling.

The works on the construction of statistical procedures of the direct modeling opened the wide possibilities on the increase of the method's effectiveness by way of decrease, literally, in several orders the volume of the operative memory of computer, as compared to that of initial modifications. This tendency permitted to apply such procedures to a solution of two-dimensional and, later, three-dimensional problems, taking into account the real properties of gases. However, by the investigation of these methods and by the substantiation of their application it is, actually, impossible to proceed without consideration of the kinetic equation describing the phenomenon to be modeled. The determination of the connection between statistical procedure and solution of the kinetic equation proves to be necessary due to a number of reasons. First, for the solution to be trusted and for the possibility of using of its results as exemplary ones keeping in mind that quite a number of typical problems was initially solved just by the methods of direct modeling, and up to the present time these solutions were not repeated with the help of other methods. Second, the establishment of correspondence between the modeling and the equation's solution permits to use the well-developed technique of numerical, both regular and statistical, methods of solution of the equations of mathematical physics for the analysis of methods under investigation and for the increase of their effectiveness. And, third, existence of such a connection permits to find some general approach to the construction of methods and to exclude the false modifications of any type.

The procedure of solution of the problems by Monte Carlo method is based on the notion of existence of the correspondence between the physical phenomenon for its description and some accidental process, of which the mathematical expectation serves as an estimate of the problem's characteristics to be found.

As a rule, the mathematical complexity of the problems in rarefied gas dynamics leads to a demarcation of a totality of computational methods into regular ones and purely statistical ones. In these cases, usually, there do not arise any doubts concerning the validity of solution of the equations and the analysis of errors is simplified due to the fact that these errors have, mainly, uniform character. The complexity of problems leads, as a rule, to the mutual interplay of approaches based on the finite differences and on the theory of probability. Thus, on one hand, statistical modeling of the gaseous flows is based on the discretization of a physical space, while, on the other hand, the different methods resort to the Monte Carlo procedures for the calculation of moments.

1.3 Connection of the Stationary Modeling with the Solution of Equation

In the process of development and investigation of the methods of statistical modeling, including that of direct modeling, it is, actually, impossible to go ahead without consideration of the kinetic equations describing the phenomenon under study. Thus, by the development of the method of direct statistical modeling connected with motion of trial trajectories, Haviland was compelled to use the Boltzmann's equation in the following iterational form:

$$\bar{\xi}\frac{df^{(k)}}{dx} = \int \left(f^{(k)}f_1^{(k-1)} - f^{(k)}f_1^{(k-1)}\right)gbdbd\varepsilon d\bar{\xi}_1. \tag{1.3.1}$$

The most evident expression for the connection between Monte Carlo methods and kinetic equation was established for linearized equation in the form (1.1.5), when for it is built up, in correspondence with the kernel of integral equation, the Uhlam–Neumann's procedure.

Equation (1.1.5) is presented in the integral form

$$\varphi(t, x, \xi) = \varphi(t_0, x - \xi(t - t_0))e^{-k(\xi)(t-t_0)}$$

$$+ \int K(\xi, \xi_1)e^{-k(\xi)(t-\tau)}\varphi(\tau, x - \xi(t - \tau), \xi_1)d\xi_1 d\tau. \tag{1.3.2}$$

When presented in more convenient form corresponding to the type of Fredholm of the 2nd kind, that equation is written down as

$$\varphi(t, y) = \psi(t, y) + \int P(t_1, y_1, t, y)\varphi(t_1, y_1)dy_1 dt_1, \tag{1.3.3}$$

where symbol y means the phase space (x, ξ).

For the molecules in the form of hard spheres well known is the appearance of its integral kernel. In dimensionless form it looks as

$$P = \frac{k_0 d^2}{\sqrt{\pi}}e^{-\xi_1^2}\left[g - \frac{2}{g}e^{\xi_1^2} - \frac{(\xi_1 g)^2}{g^2}\right]e^{-k_0 d^2 t}\frac{1}{\sqrt{\pi}}\int ge^{\xi_1^2}d\xi_1. \tag{1.3.4}$$

Correlated with Eq. (1.3.2) is the uniform Markov's chain with initial distribution corresponding to the initial distribution function

$$\Psi(t, y) = \varphi[t_0, y - \xi(t - t_0)]e^{-k(\xi)(t-t_0)},$$

and with the matrix of transition corresponding to the kernel (1.3.4).

In the case at consideration the probability of sequence

$$(t_1, y_1) \rightarrow (t_2, y_2) \rightarrow \cdots (t_l, y_l),$$

consisting of l dispersions, is equal to

$$\Psi(t_1, y_1)P(t_1, y_1 \rightarrow t_2, y_2)$$
$$P(t_2, y_2 \rightarrow t_3, y_3) \cdots P(t_{l-1}, y_{l-1} \rightarrow t_l, y_l)dt, \quad dy \cdots dt_l, dy_l,$$

and the mathematical expectation for some accidental quantity

$$X = \sum_{i=1}^{l} \xi(t_i, y_i),$$

will be equal to the function of the solution of initial equation $(\varphi, \Psi) = M[X]$.

Solution of the nonlinear kinetic equation with the help of method of stationary statistical modeling is obtained, actually, by the method of iterations. In the frame of each kth iteration one obtains the linear integral equation which is solved using Monte Carlo method. Thus, for example, for the model Krook's equation this solution looks as

$$f^{(k+1)} = f_n e^{-\int v^{(k)} dt} + \int \frac{f_0^{(k)}}{n^{(k)}} e^{-\int v^{(k)} dt} f^{(k+1)} d\xi dt. \qquad (1.3.5)$$

Within each iteration calculated are the macroparameters entering the initial equation, $n^{k+1}, u^{k+1}, T^{k+1}, v^{k+1}$, and after that on proceeds to the next iteration. If the method of successive approximations converges, then the transition from one iteration to another will, after all, lead to the solution of a kinetic equation.

1.4   Construction of the Method of Direct Statistical Modeling

Let us consider, for simplification, the uniform Boltzmann's equation

$$\frac{\partial f}{\partial t} = J(f, f). \qquad (1.4.1)$$

This equation will be solved by the method of successive approximations. Such a method will be chosen in Euler' form

$$\begin{aligned} f^0 &= f(0), \\ f^1 &= f^0 + J(f^0, f^0)\Delta t, \\ f^2 &= f^1 + J(f^1, f^1)\Delta t, \\ f^n &= f^{n-1} + J(f^{n-1}, f^{n-1})\Delta t. \end{aligned} \qquad (1.4.2)$$

Let us introduce into the area of solution the computational Lagrangian network, that is, let us represent the distribution function by some system of N particles, each of which is characterized by the velocity of motion ξ_i. The set of Eq. (1.4.2) will be solved with the help of the method of fractional steps, which means that at the temporal half-step one should consider the change of the internal state of the set as a whole. Following this way, the calculation process might be described as a sequence of stages: first, at the Lagrangian stage of calculations in each approximation on time one determines the transfer of particles for this particular case:

$$f_i^n = f_i^{n-1/2}, \quad \Delta\xi f(t, \xi_i) = f\left(t - \frac{\Delta t}{2}, \xi_i\right)\Delta\xi, \tag{1.4.3}$$

where f_i is the part of distribution function, which is approximated by the particles with velocities in vicinity of ξ_i, while the Eulerian stage, similarly to what is done in semiregular methods, might be estimated with the help of Monte Carlo methods. Thus, from the form of relaxation stage

$$\Delta\xi f^{n+1/2}(t + \Delta t, \xi_i) = \frac{\Delta t}{2}T_i^n\Delta\xi - v^n(t)\frac{\Delta t}{2}f^n(t, \xi_i)\Delta\xi, \tag{1.4.4}$$

it is seen that the part of molecules, proportional to the frequency of collisions at the preceding step and to the size of temporal step, does not alter velocities, while another part of molecules alters velocities in correspondence with the quantity $J^n(t, \xi_i)$. In particular, the calculation of right-hand side of Eq. (1.4.4) might be conducted in such a way. Considering the quantity $v^n(t)$ and Lagrangian network (that is, taking into account the number of particles approximating the distribution function), at the nth step one is able to determine the length of time spent for one collision, and in correspondence with probability of collision $\xi(g)$ one chooses the pair of molecules participating in the collision. The number of colliding pairs is chosen in such a way that the total length of time spent on collisions would not surpass the temporal step. The rest totality of particles does not alter velocities during this temporal step. Thus, after the relaxation stage the part of particles alters the velocities, while another part does not lead to alteration, though Lagrangian network as a whole will be changed. After that, once again, is realized step in time, and the process goes on in correspondence with a set (1.4.2), up to the convergence of successive approximations to the solution of Boltzmann's equation (1.4.1). It is possible to see that the process of solution of the kinetic equation, described above, is similar to the Bird's procedure of direct modeling.

The simple reasonings like given above permit, as it will be seen later, to estimate the errors of methods and to construct the schemes of solution.

Chapter 2

Development of the Numerical Methods of Solution of the Linear Kinetic Equations

2.1 The Perfection of VGK Method (Vlasov, Gorelov, Kogan)

Proposed in Ref. 97 was the procedure of solution of the linearized Boltzmann's equation by Monte Carlo method. Using this procedure was obtained an exact solution of the number of problems in rarefied gas dynamics (Kogan, Vlasov, Gorelov, Khlopkov[97,98]). In Ref. 97, this method was used for pseudo-Maxwellian molecules, for which the collisional cross section is inversely proportional to the relative velocity of colliding particles, $\sigma = \sigma_0/g$. Conducted in Ref. 98 was the study of the convergence and accuracy of the method by its extension to the other laws of molecular interaction.

Under consideration is the weakly perturbed flow. In this case, the distribution is slightly different from some characteristical Maxwellian distribution function, that is

$$f = f_{00}(1 + \varphi),$$
$$f_{00} = n_0 \left(\frac{m}{2\pi k T_0} \right)^{3/2} \exp \left(\frac{m}{2k T_0} \xi^2 \right), \quad \varphi \ll 1. \tag{2.1.1}$$

Here m is mass of a molecule, k — constant of Boltzmann, ξ — velocity of a molecule, φ — the small addition, the square of which is negligibly small.

The process of transfer might be considered as a uniform chain of Markov, the links of which happen to be the positions of a particle immediately before the collision. Since one is considering small additions to the distribution function and its moments, statistical errors might be of the order of quantities to be looked for. For this reason, the wandering of the trial molecule is gambled over the equilibrium distribution function f_{00}, and subtracted from the total transfer is that obtained for the absolutely Maxwellian distribution. In this case, the mean transfer of some molecular property through some chosen plane is equal to

$$\Psi = \frac{1}{N} \sum_{\alpha=1}^{N} \left(\frac{W}{W_0} - 1 \right) \sum_{\beta(\alpha)} \psi_{\alpha\beta}, \tag{2.1.2}$$

30

where N is the number of trajectories, W and W_0 — the probabilities of a flight of particle along the trajectory α for, correspondingly, perturbed and unperturbed flows, $\beta(\alpha)$ — the number of crossings of the plane chosen by this trajectory.

The molecule's wandering is realized in following ways.

1. Gambled at one of the boundaries are velocities of the molecule flying into the area under consideration with the density of probability proportional to $f_{00}(\vec{\xi}) \cdot (\vec{\xi}, \vec{n})$.

2. Flight of the temporal length τ without collisions and collision within element $d\tau$ is gambled with probability

$$d\tau \exp\left(-\int \left[\int f_{00} g\sigma d\vec{\xi}_1\right] d\tau\right) \int f_{00} g\sigma d\vec{\xi}_1.$$

3. Velocity of the molecule-partner is gambled with the density of probability $g\sigma f_{00}(\vec{\xi}_1)$.

4. Collision is gambled in accordance with the law of interaction of the molecules involved, and further on, beginning with point 2 and until the molecule leaves the area of flow.

The relation W/W_0 is found according to formula

$$\frac{W}{W_0} = \frac{F_w}{F_{w_0}} \frac{P}{P_0} \frac{F}{F_0} \frac{Q}{Q_0} \cdots, \tag{2.1.3}$$

where F_w and F_{w_0} are probabilities of the molecule's flight out of the boundary, correspondingly, for perturbed and unperturbed flows; P and P_0 are probabilities of the free flight during the time τ; F and F_0 are probabilities of collision within the element $d\tau$ with a molecule flying with a velocity $\vec{\xi}_1$; Q and Q_0 — probabilities of collision with the prescribed aiming distance. Using the problem's linearity, expression (2.1.3) might be presented in the form of sum

1. $$F_w = \frac{(\vec{\xi}_1, n) f_0 (1 + \varphi_w) d\vec{\xi}}{N_u}, \quad F_{w_0} = \frac{(\vec{\xi}_1, n) f_0 d\vec{\xi}}{N_0},$$

$$N_u = \int (\vec{\xi}_1, \vec{n}) f d\vec{\xi}, \quad N_0 = \int (\vec{\xi}_1, \vec{n}) f_0 d\vec{\xi}.$$

2. $$P = \exp\left(-\iint f_1 g\sigma d\vec{\xi}_1 d\tau\right) d\tau \int f_1 g\sigma d\vec{\xi}_1$$

$$= \exp\left(-\iint f_{01}(1 + \varphi_1) g\sigma d\vec{\xi}_1 d\tau\right) d\tau \int f_{01}(1 + \varphi_1) g\sigma d\vec{\xi}_1,$$

with the notations: $K_0 = \int f_0 g\sigma d\vec{\xi}_1$, $K_0 k = \int f_0 \varphi g\sigma d\vec{\xi}_1$.

After that, using the fact that $k \ll 1$, one obtains

$$P = \exp\left(-K_0\tau - K_0 \int k d\tau\right) d\tau(K_0 + K_0 k)$$

$$= \exp\left(-K_0\tau\right)\left(1 - K_0 \int k d\tau\right) d\tau K_0(1 + k).$$

Similarly to that, for $P_0 = \exp(-K_0\tau)d\tau K_0$ one obtains

3. $$F = \frac{f_1 g\sigma\xi_0}{\int f_1 g\sigma d\vec{\xi}_1}, \quad \frac{f_{01}(1 + \varphi_1)g\sigma d\vec{\xi}_1}{FK_0(1 + k)}, \quad F_0 = \frac{f_{01} g\sigma\xi_1}{K_0}.$$

4. Quantities Q and Q_0 are equal both for equilibrium and genuine trajectories. Upon inserting the values of probabilities obtained into expression (2.1.3) one finds

$$\frac{W}{W_0} = \frac{N_0}{N_u}(1 + \varphi_w)\left(1 - K_0 \int k d\tau\right)(1 + \varphi_1) \cdots$$

$$= \frac{N_0}{N}\left(1 + \varphi_w - K_0 \int k d\tau + \varphi_1 + \cdots\right). \qquad (2.1.4)$$

Taking into account that from the surface of a boundary flies in the number SN_u of molecules, one obtains, finally

$$\Psi = \frac{SN_0}{N} \sum_{\alpha=1}^{N}\left(\varphi_w - K_0 \int k d\tau + \varphi_1 + \cdots\right)\sum_{\beta(\alpha)} \psi\alpha\beta, \qquad (2.1.5)$$

where S is the area of bounding surface.

As it follows from above reasonings, for the gambling of a trial molecule it would be necessary to know the distribution function of the field particles. Therefore, the problem is solved by the method of successive approximations. For the initial approximation one might take any function, for example — zero. The process of calculation goes on up to the point when $\varphi^{(n)} \approx \varphi^{(n-1)}$ with the prescribed degree of accuracy. The process of calculations according to the methodics described is, to some degree, equivalent to a solution of Boltzmann's equation using the following iterational scheme:

$$\frac{df^{(n)}}{dt} = -f^{(n)} J_2^{(n-1)} + J_1^{(n-1)}.$$

Actually, one has

$$f = f_w \exp\left(-\int J_2 d\tau\right) + \int J_1 \exp\left(-\int J_2 d\tau\right) d\tau, \qquad (2.1.6)$$

where f_w is an initially prescribed function.

Upon linearization of this equation one obtains

$$J_2 = \int fg\sigma d\vec{\xi}_1 = K_0(1+k),$$

$$J_1 = \int f'f_1'g\sigma d\vec{\xi}_1 + \int f_0 f_{01}(1+\varphi'+\varphi_1')g\sigma d\vec{\xi}_1$$

$$= f_0 K + \int f_0 f_{01}(\varphi'+\varphi_1')g\sigma d\vec{\xi}_1. \tag{2.1.7}$$

As soon as in the last term of the expression for J_1 the primed quantities will be substituted by unprimed ones in accordance with the conservation formula

$$\vec{\xi}' = \vec{\xi} + \vec{n}(\vec{g}, \vec{n}),$$
$$\vec{\xi}_1' = \vec{\xi} - \vec{n}(\vec{g}, \vec{n}),$$

one obtains

$$J_1 = f_0 K_0 + \int f_0 f_{01}\varphi[\vec{\xi}_1\vec{\xi}]g\sigma d\vec{\xi}_1. \tag{2.1.8}$$

After the substitution of values J_1 and J_2 into Eq. (2.1.6) one finds that the functional looked for will appear as ($\Psi = \Psi_0 + \psi, \psi \ll 1$)

$$\Psi = \int f_w \exp\left(-\int J_2 d\tau\right)\psi d\vec{\xi} + \iiint f'f_1' \exp\left(-\int J_2 d\tau\right)\psi g\sigma d\vec{\xi}_1 d\vec{\xi} d\tau.$$

Then, after the change of variables, $\xi', \xi_1' \to \xi, \xi_1$, and after the use of theorem on conservation of the phase volume, one obtains

$$\Psi = \int f_w \exp\left(-\int J_2 d\tau\right)\psi d\vec{\xi} + \iiint f^{(n)} f_1 \exp\left(-\int J_2 d\tau\right)\psi' g\sigma d\vec{\xi}_1 d\vec{\xi} d\tau.$$

In the iterational scheme (2.1.6) quantity Ψ will appear as

$$\Psi^{(n)} = \int f_w \exp\left(-\int J_2^{(n-1)}d\tau\right)\psi d\vec{\xi}$$

$$+ \iiint f^{(n)} f_1^{(n-1)} \exp\left(-\int J_2^{(n-1)}d\tau\right)\psi' g\sigma d\vec{\xi}_1 d\vec{\xi} d\tau. \tag{2.1.9}$$

And, after insertion of the values (2.1.7) and (2.1.8) into (2.1.9) one obtains

$$\Psi^{(n)} = \int\left(\varphi_w - K_0\int k^{(n-1)}d\tau\right)f_0\psi d\vec{\xi}$$

$$+ \iiint\left(\varphi^{(n)} + \varphi^{(n-1)} - K_0\int k'd\tau\right)f_{01}\psi' g\sigma d\vec{\xi}_1 d\vec{\xi} d\tau. \tag{2.1.10}$$

As it is seen from such a form of presentation, expression (2.1.5) is nothing but the mathematical expectation of the functional (2.1.10).

As it was noted above, for the computation of functional to be found in each point of the physical space it would be necessary to memorize the distribution function. Moreover, quite essential difficulties arise in connection with the computation of three-dimensional integral $K_0 k = \int f_{01} \varphi_1 g \sigma d\vec{\xi}_1$, which is needed to be calculated several times within the frame of a single trajectory. For this reason, for the diminishment of the computational time before the next iteration these integrals were calculated separately and were listed in the table.

The problem is essentially simplified in the case of pseudo-Maxwellian molecules, when $\sigma = \sigma_0/g$. In this case $K_0 = n_0 \sigma_0$ and $K_0 k = n_0 \sigma_0 \nu$, where ν is an additional density, which is independent of velocities. Because of that one is getting rid of the necessity of calculation of the integrals. Therefore, due to the relative simplicity, the numerical study with the help of this method was conducted on the solution of the problem of computation of the transfer coefficients for pseudo-Maxwellian molecules.

The problem was formulated as follows.[97] A certain layer was cut from the infinite area having the prescribed gradients of velocity and temperature, of which the thickness is of the order of the mean free path's length (see Fig. 2.1).

The gradients $\frac{\partial T}{\partial x}$ and $\frac{\partial u}{\partial x}$ should comply to the condition that at the distance of mean free path's length the macroparameters T and u endure just insignificant variations.

Formula (2.1.5) is used for computation of the stress tensor and of the flow of energy across the plane $x = 0$, and according to the well-known expressions

$$P_{xy} = \mu \frac{\partial u}{\partial x}, \quad q_x = -\lambda \frac{\partial T}{\partial x},$$

are found the coefficients of both viscosity and heat conductivity. The numerical experiment was conducted to determine the dependence of the computational accuracy of the number of trajectories gambled, on the number of cells for velocities and on the length of axes for velocities. It was found that for the computation

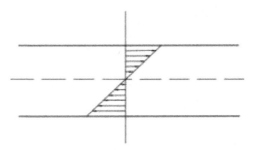

Fig. 2.1

of stress tensor one needs to have axes of the length of 2.5 thermal speeds. The further increase of the length of axes does not affect the accuracy of results. For the computation of a heat flow, it would be necessary to cut-off not less than 2.8 thermal speeds.

The deterioration of accuracy in the case of a heat transfer might be explained by the fact that q_x is representing the moment, order of which is higher than that of P_{xy}.

Considered above was the behavior of the integral characteristics of flow, but, moreover, the distribution function was studied, too.[98] One should note the fact of good agreement of the values obtained by the most probable velocities, as well as complete agreement with a change of sign, which is of great importance for the computation of the integral characteristics of flow.

The results of the study of a scheme for pseudo-Maxwellian molecules were extended to a molecular model in the form of hard spheres. The coefficients of viscosity and heat conductivity were calculated and compared with their theoretical values.

Thus, taking into account the results of Refs. 97, 98 one could make a conclusion that the method described might be applied for the case of molecules with arbitrarily cut-off potentials of interactions, as well as for the case of gaseous mixture.

2.2 Modification of the Vlasov's Method for the Solution of Linear Problems

Considered below is the method of mathematical modeling of the linear processes in the rarefied gas. In distinction of the method of the preceding paragraph, the present method permits to realize the computations without memorizing of the distribution function, which quality permits, in its turn, to reduce essentially the operative memory of the computer (Vlasov, Khlopkov[99]). This method presents itself as a natural extension of Vlasov's method[35] to the linear problems.

The essence of the method is reduced to the following. The field of a flow is divided in cells. As soon as the trial molecule is brought into the cell, within which occurs the collision, it would be necessary to know the distribution function of the field molecules within that cell in order to find the velocity of the trial one after collision. In order to avoid keeping in the computer's memory whole distribution function such a procedure is proposed. Memorized in each cell is the velocity of trial molecule with the probability proportional to Δt, that is, to the length of time of staying of the trial molecule within that cell. Since the frequency of the visiting this cell by the trial molecule is proportional to $\xi f(\xi)$, the frequency of its memorizing is proportional to $\xi f(\xi) \Delta t$. But $\Delta t \sim 1/\xi$, which means that the frequency of memorizing is proportional to $f(\xi)$. Thus, by the collision of a trial molecule in the chosen cell the part of the velocity of the molecule-partner is played by the speed of filling of that cell with a probability proportional to Δt. The computational scheme of that method is essentially simplified if one considers the pseudo-Maxwellian molecules. In this case, the frequency of collisions is equal to $N = \int f_1 g\sigma d\vec{\xi} = \sigma_0 n$. The probability of collision within the chosen cell is equal to $\sigma_0 n \Delta t$. The probability of the flight of the molecule into that particular cell is proportional to $\xi f(\xi)$. And, after all, the probability of collision of the trial molecule within that cell is proportional to $\xi f(\xi)\sigma_n \Delta t - \sigma_0 nf - f(\xi)$.

As it follows from the preceding discussion, the computation might be carried out according to such a scheme: in the cell, where the collision takes place, one should memorize the velocity with which the particle flies into that cell, while the computation of collision should be realized with a velocity stored in memory as a result of the preceding collision.

For a linear case, the structure of method looks as follows. Prescribed at the boundary is the distribution function of the molecules flying in, and this function is just slightly different from an equilibrium one. The macroscopic parameters are also slightly different from the characteristical equilibrium values:

$$T = T_0(1 + \varepsilon\theta), \quad n = n_0(1 + \varepsilon v), \quad \varepsilon \ll 1.$$

The gambling of trajectories is realized on the equilibrium distribution function:

$$f_0 = n_0 \left(\frac{m}{2\pi k T_0} \right)^{3/2} \exp\left(\frac{m}{2k T_0}\xi^2 \right),$$

whereas the part of genuine parameters of the trajectory is played by the parameters of the equilibrium one plus additions of the order of ε:

$$\vec{v} = \vec{\xi} + \varepsilon\vec{\zeta}, \quad v = \tau + \varepsilon\sigma\tau,$$

where $\vec{v}$ is velocity and v — the time of free path for genuine trajectory, τ — the same time for equilibrium one.

1. The velocity of a trial molecule, which flew into the field of flow, is gambled according to the boundary conditions over the equilibrium distribution function, while the value of a genuine velocity is determined in accordance with a formula $\vec{v} = \vec{\xi} + \varepsilon\vec{\zeta}$, where $\vec{\xi}$ depends on the parameters at the boundary.
2. The free path, where R is an accidental number which is uniformly distributed between 0 and 1.
3. The collision. Before the collision within each cell is memorized the value of velocity with which the molecule entered the cell, where the collision occurs,

while the collision is described by a formula

$$\vec{v}' = \frac{1}{2}\left(\vec{v} + \vec{v}_1 + \vec{n}G\right), \quad G = |\vec{v} - \vec{v}_1|,$$

$$\vec{\zeta} = \frac{1}{2}\left(\vec{\zeta}_1 + \vec{\zeta}_1 + \vec{n}\frac{(\vec{g}, \vec{g}_1)}{g}\right), \quad g_1 = |\vec{\zeta} - \vec{\zeta}_1|,$$

where $\vec{v}_1$ is a velocity resulted from a preceding collision. From here one obtains separately for ξ and ζ:

$$\vec{\xi}' = \frac{1}{2}(\vec{\xi} + \vec{\xi}_1 + \vec{n}g), \quad g = |\vec{\xi} - \vec{\xi}_1|.$$

Thus the molecule will wander up to that moment when it leaves the area of flow.

Let $\psi(v)$ be a certain molecular property, which permits to calculate macroparameters Ψ with the help of formula

$$\Psi = \frac{\sum \psi \Delta t}{\Delta S N / N_u},$$

where Δt is a length of time of stay of a trial molecule within the cell, N is a number of trajectories, and N_u — a number of molecules flying in from the boundary.

To prevent the clogging of the small deviations from the equilibrium problems, which are to be found, by the statistical errors, it will be necessary to decompose the values of Ψ in Ψ_0 and $\delta\Psi$ and to subtract Ψ_0. After that

$$\delta\Psi = \frac{\Psi - \Psi_0}{\varepsilon} = \frac{1}{\varepsilon \Delta S}\left(\frac{N_u}{N}\sum \psi \Delta t_0 - \frac{N_0}{N}\sum \psi \Delta t_0\right)$$

$$= \frac{N_0}{N\Delta S}\left[SN\sum \psi_0 \Delta t_0 + \sum \delta\Psi \Delta t_0\right]$$

or, finally,

$$\delta\Psi = \frac{N_0}{N\Delta S}\sum[\psi_0(\delta N \Delta t_0 + \delta t) + \delta\psi \Delta t_0].$$

Here N_0 symbolizes the number of molecules flying into the cell in equilibrium state:

$$\Delta t = \Delta t_0 + \varepsilon \Delta t, \quad N = N_0 + \varepsilon \delta N.$$

The method described was tested in application to the Couette's problem on the flow between parallel plates with equal temperatures, which are moving, one with respect to another, with rather small velocities.

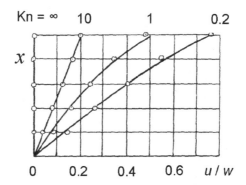

Fig. 2.2   Profil of the velocity in the Couette flow.

Prescribed at the boundary is the diffuse reflection of molecules:

$$f_\pm = n_0 \left(\frac{m}{2\pi k T_0} \right)^{3/2} \exp\left(-\frac{m}{2kT_0} [(\xi_x \pm \varepsilon w)^2 + \xi_y^2 + \xi_z^2] \right).$$

The area of flow is divided in cells. Used in the present work was division in 20 cells.

Presently we have $\delta N = 0$, and thus the velocity to be found is equal to

$$U = \frac{N_0}{N\Delta S} \sum (\xi_x \delta t + \zeta_x \Delta t_0).$$

In the present case one could give the approximate estimate of the error:

$$\delta = 5\% + \sqrt{\frac{D}{N}}.$$

The method has revealed a good convergence and an agreement with the results obtained earlier for Boltzmann's equation and described in Ref. 97.

Shown in Fig. 2.2 is the dependence of the behavior of the flow's velocity, as it is related to the wall's velocity, on the Knudsen's number. The points in this figure correspond to the data obtained in Ref. 97, while the solid lines correspond to the results of the present work.[99]

2.3   Method of Solution of the Linearized Boltzmann's Equation

Developed in the preceding sections of the present chapter were the methods based on the use of trial particles, and in spite of the fact that these methods revealed their effectiveness by the solution of the number of practical problems (see Chap. 5), they demand the division of physical area in cells (and the first method demands, moreover, the division of a phase area); such a division is not always necessary by the numerical solution of the classical linear equations of mathematical physics. Therefore, seems to be quite natural the attempt to find some method of solution of the linear problems which would not demand any division of the area in cells (see Khlopkov[36]).

Let us consider the flows weakly different from the equilibrium ones. The distribution function for such a case might be presented in the form

$$f = f_0(1 + \varphi), \quad \varphi \ll 1, \tag{2.3.1}$$

where

$$f_0 = n_0 \left(\frac{m}{2\pi k T_0} \right)^{3/2} \exp \left(\frac{m}{2k T_0} \xi^2 \right),$$

and the Boltzmann's equation is linearized, assuming the form

$$\xi_x \frac{\partial \varphi}{\partial x} + \xi_y \frac{\partial \varphi}{\partial y} + \xi_z \frac{\partial \varphi}{\partial z} = -\varphi k(v) + \int K(\xi, \xi_1) \varphi_1 d\xi_1, \tag{2.3.2}$$

where $\xi = \xi \sqrt{\frac{m}{2kT}}$.

For the molecules, which interact according the law of hard spheres, the analytical expressions appear as

$$k = a \int g e^{-v_1^2} dv_1, \tag{2.3.3}$$

$$K = B e^{-v^2} \left(g - \frac{2}{g} e^{D^2} \right), \tag{2.3.4}$$

where $g = v_1 - v$, while a and B are constant factors.

Upon carrying out the integration of Eq. (2.3.2) along the trajectories one comes to the linear integral equation

$$\varphi = \varphi_r + \int K \varphi_1 dv_1 dl, \tag{2.3.5}$$

where

$$\varphi_r = \varphi_w e^{-\frac{k}{v}(l-l_0)}, \tag{2.3.6}$$

$$K = \frac{1}{v} K e^{-\frac{k}{v}(l_1-l_0)}. \tag{2.3.7}$$

Applicable to Eq. (2.3.5) is the statistical procedure of Uhlam and Neumann. In the general case calculation by the Monte Carlo method is reduced to that of integrals. The part of these integrals is played by the mathematical expectations of casual values, which are used as estimates. In other words, one performs the estimation of the integral of Lebesque–Stilltjess'es type on some probabilistical measure

$$l = \int \psi(x) u dx, \tag{2.3.8}$$

with the help of a mean arithmetical quantity over the number of tests:

$$\frac{1}{N} \sum_{i=1}^{N} \psi(x_i). \qquad (2.3.9)$$

In the case of integral equations, the integral is nothing more nor less than the functional of the solution of equation

$$\varphi(x) = \varphi_r(x) + \int K(x, y)\varphi(y)dy, \qquad (2.3.10)$$

where D is an area of the S-dimensional Euclidian space $x \in D$, $y \in D$; $\varphi(x)$ and $\varphi_r(x)$ are the functions determined within D, the kernel $K(x, y)$ is determined on the Cartesian product of D by itself, and the integral in the last equation is appreciated in Lebesque's sense. As it is easy to see, Eq. (2.3.5) is just a particular case of the general linear integral equation (2.3.10). The solution of such an equation is given by the convergible Neumann's series:

$$\varphi(x) = \sum_{l=0}^{\infty} \psi_i(x),$$

where $\psi_i(x) = \int_D K(y, x)\psi_{i-1}(y)dy$.

In this case, the functional we are looking for is presented in the form of the sum of i-fold integrals:

$$I = (\varphi, h) = \sum_{i=0}^{\infty} \int_{R_i} \varphi_r(x_0)K(x_0, x_1) \cdots K(x_{i-1}, x_i)h(x_i)d\theta_i,$$

where $R_i = \underbrace{D \times D \times D \cdots \times D}_{i}$, $R_0 = D$, $d\theta_0 = dx_0$, $d\theta_i = dx_0 dx_1 \cdots dx_i$.

Usually, Eq. (2.3.10) is associated with a uniform Markov's chain, prescribed by the density of initial distribution and the transitional density $p(x \to y)$. The selective trajectory of that chain is built up in correspondence with the initial and transitional densities of probability. Then the estimate (2.3.9) of the functional we are looking for will appear as

$$\psi = \sum_{i=0}^{k} Q_i(x_0, \ldots, x_i)h(x_i),$$

where

$$Q_0 = \frac{\varphi_r(x_0)}{\pi(x_0)},$$

$$Q_i = Q_0 \frac{k(x_0, x_1)}{p(x_0 \to x_1)} \cdots \frac{k(x_{i-1}, x_i)}{p(x_{i-1} \to x_i)},$$

and where k is the number of collisions on the trajectory.

Thus, as soon as the analytical expression for the kernel is known, it would be possible to construct the simple algorithm for computation of the various functionals of solution of the integral equation.

In the process of a numerical solution of the equation, the part of the Markov's chain is played by the trajectories of molecules with the equilibrium distribution function. In this case one might use the simple process of drawing of the trajectories, which are close to the genuine ones, and such a practice leads to the essential diminishment of a dispersion. The process of modeling of the accidental quantity happens to be completely equivalent to the method of successive approximations, and, therefore, the solution built up by Monte Carlo method converges to the solution of Boltzmann's equation.

The approbation of that method was realized on the problem of calculation of the transfer coefficients. Cut out of the area of flow with the prescribed values of gradients of the macroparameters is the gaseous layer, of which the thickness is of the order of the molecular mean free path. It is assumed that both at the upper and lower boundaries one has the distribution of Navier–Stokes with the prescribed values of velocities and their gradients. There exists an exact solution of such a problem, obtained, for example, by Chapman–Enskog's method.

Thus, the molecules flying in from the upper and lower boundaries will have the known distribution function, and the transfer of a molecular property is computed across the symmetry plane. The method was checked on the computation of the viscosity coefficients for the molecules in the form of hard spheres. In that case, the functional

$$p_{xy} = -\mu \frac{\partial u}{\partial x} = \int f_0 \phi \xi_x \xi_y d\vec{\xi}$$

is calculated with the help of a formula of the mean arithmetical value. As an example of Markov's chain is chosen the wandering of molecules on the absolutely equilibrium distribution function f_0.

Presented in Fig. 2.3 is the dependence of velocity on the number of trajectories N (in thousands) for the molecules — hard spheres. In view of the fact that in the present case neither physical nor speedy space is not divided in cells, and that there are not any limitational assumptions, the statistical error might be reduced to zero by the increase of the number of tests. The systematical error which is obtained depends solely on the quality of the counter of accidental numbers:

$$\delta \approx 3\% + \sqrt{\frac{D}{N}}.$$

The numerical computation according to the methodics developed and described above possesses the number of the advantages, as compared with other

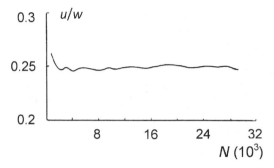

Fig. 2.3   Couette flow.

methods. It permits to solve the linearized Boltzmann's equation not only without memorizing the distribution function, but also without memorizing the values of macroparameters, due to the fact that as soon as the distribution function at the boundaries is prescribed, then the solution in the first approximation would be found.

Chapter 3

Methods of Solution of the Nonlinear Problems in Rarefied Gas Dynamics

3.1 Method of Solution of the Model Equation Based on a Stationary Modeling

Described in this section is the method of solution of linear and nonlinear problems of the rarefied gas dynamics which would permit to manage without memorizing the distribution function, when the part of basic equation is played by a simplified model of Boltzmann's equation presented in the form (see Kogan[8])

$$\frac{df}{dt} = An(f_0 - f).$$
(3.1.1)

The presentation is realized by Khlopkov.[39]

The main distinction of the model equation from that of Boltzmann consists in the fact that model equation is based on assumption that the distribution function of molecules after the collision becomes the most probable with prescribed values of the numbers of colliding particles, of their momentum and energy. Acquaintance with the data on macroparameters completely determinates the trajectory of a trial particle, since the molecular velocities after the collision are distributed in accordance with the function

$$f_0 = n \left(\frac{m}{2\pi kT}\right)^{3/2} e^{-\frac{m}{2kT}(\xi - u)^2}.$$
(3.1.2)

In the case of linearity computed are just the small additions to the equilibrium parameters, which means that, like is was in the preceding case, the trajectories are gambled on the absolutely equilibrium distribution function f_{00}, while the transfer of a molecular property is computed with the weight W/W_0, and subtracted from the complete transfer is one, realized on the absolutely equilibrium distribution function,

$$f_{00} = n_0 \left(\frac{m}{2\pi kT_0}\right)^{3/2} e^{-\frac{m}{2kT_0}\xi^2}.$$
(3.1.3)

The trial particle's trajectory is constructed according to the following scheme:

(1) The molecules fly out the boundary with a density of probability

$$\frac{f_{00}(\bar{\xi}, \bar{n})}{N_0}.$$

(2) The duration τ of the collisionless flight and collision within the element $d\tau$ is gambled with a probability $e^{-An_0\tau}An_0 d\tau$.

(3) The velocities after the collision are distributed with a probability f_{00}/n_0, and process goes on, beginning with p. 2, up to the moment when the molecule leaves the area of flow.

The ratio of probabilities W/W_0 is built up like it was done in p. 2 of Sec. 2.2 of the preceding chapter. In this case one obtains

$$\frac{W}{W_0} = \frac{N}{N_0}\left(1 + \varphi_w - An_0 \int v d\tau + v + \varphi_r + \cdots\right), \qquad (3.1.4)$$

where φ_r is determined from the formula

$$f_0 = f_{00}(1 + \varphi_0).$$

The macroparameters $\bar{\Psi}$ correspond to the expression

$$\bar{\Psi} = \frac{1}{N}\sum_{\alpha=1}^{N}\left(\frac{W}{W_0} - 1\right)\sum_{\alpha}\sum_{\beta(\alpha)}\Psi_{\alpha\beta}$$

$$= \frac{N_0}{N}\sum\left(\phi_w - An_0\int v d\tau + v + \phi + \cdots\right). \qquad (3.1.5)$$

The distribution function is not contained in expression (3.1.5) and, therefore, it would be sufficient to know those macroparameters which enter the formula for φ_A.

It is possible to show that computation in accordance to the methodics described above is similar to the solution with the help of the following iterational scheme:

$$f^{(i)} = f_w e^{-\int An^{(i+1)}d\tau} + \int An^{(i-1)} f_0^{(i-1)} e^{-\int An^{(i-1)}d\tau} d\tau, \qquad (3.1.6)$$

with the functional $\bar{\Psi} = \bar{\Psi}_0 - \bar{\psi}$ corresponding to expression

$$\Psi^{(i)} = \left(f_w e^{-\int An^{(i-1)}d\tau}, \psi\right).$$

Linearization of this equation gives a result:

$$\bar{\Psi}^{(i)} = \left(e^{-An_0\tau}f_{00}(\varphi_w - An_0 \times \gamma\,d\tau), \psi + \int\left(An_0 e^{-An_0\tau}f_{00}\right.\right.$$

$$\left.\left. + \int\left(\gamma^{(i-1)} + \varphi^{(i-1)} - An_0\int v^{(i-1)}d\tau\right)d\tau, \psi\right)\right). \qquad (3.1.7)$$

As it is evident from such a presentation, the expression in (3.1.5) is just a mean arithmetic estimation of the mathematical expectation. The approbation of this method was realized on the linear problem of the Couette flow. Tested on that comparatively simple flow were almost all the known methods of solution of the Boltzmann equation. Considered is a flow between plane, parallel, and infinitely long plates by the various Knudsen numbers, small relative velocities $W_\pm$ and small ratio of temperatures $T_{W_\pm}$ of these plates. With this assumptions, the problem is linearized and, moreover, the problem as a whole might be splitted into two minor problems: the problem of a pure displacement by $T_{W_+} = T_{W_-}$, for which $\gamma = \tau = 0$, and the problem of a heat transfer, for which $U = 0$. Here γ and τ are just dimensionless additions to density, n_0 and T_0, correspondingly.

When presented in dimensionless form, Eq. (3.1.1) looks as

$$\frac{v_x}{d}\frac{d\varphi}{d\tau} = -\phi + \gamma + 2v_y u_y + \left(v^2 - \frac{3}{2}\right)\tau, \qquad (3.1.8)$$

where $v = \xi\sqrt{n_0}$, $\alpha = An_0d\sqrt{n_0}$ — the quantity inversely proportional to the Knudsen number. Taking into account that $A = \frac{kT}{\mu}$, where μ for Maxwellian molecules is equal to $\frac{2kT}{3}$ and $\lambda = \frac{1}{n_03_0\sqrt{n_0}}$, one obtains $d = \frac{n_03_0d\sqrt{n}}{2} = \frac{1}{2kn}$.

The field of flow between the plates is splitted into 10 cells, and calculated for interior of each cell are the additions to the physical macroparameters of the flow.

The problems of a pure displacement and of a heat transfer were solved separately one of the other (the temperature's profiles are shown in Fig. 3.1 and, practically, do not differ of those of exact numerical solution by Willis).

The solution obtained differs from the exact numerical solution of the model equation in something of the order of 5%, though the statistical error by $N > 50,000$ does not exceed 1%.

The scheme proposed is valid in the nonlinear case, too.

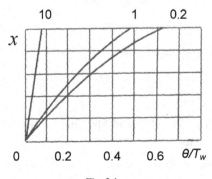

Fig. 3.1

The computation is carried out by the method of successive approximations, and used for each nth approximation are the values of n, u, and T, taken from the $(n-1)$th approximation.

The trajectory of a trial molecule is built up in following ways.

(1) At the flow's boundary the molecular velocity is distributed accordingly to the distribution function f_0.

(2) The free path during the time τ is realized with a probability

$$e^{-\int An^{(n-1)}d\tau}.$$

(3) The velocities after the collision are distributed in accordance with $f_0^{(n-1)}$, and so on, beginning with the item 2.

The parameters $\bar{\Psi} = (n,\ u,\ T)$ we are looking for might be found in accordance with the formulae $\bar{\Psi} = \frac{1}{N}\sum \psi$.

The process of computations with the help of methodics described above is more or less equivalent to the solution of equation in accordance to the following scheme:

$$f^{(n)} = f_w e^{-\int An^{(n-1)}d\tau} + \int A_n^n f_0^{(n-1)} e^{-\int An^{(n-1)}d\tau} d\tau.$$

And really, since $n = \int f_1 d\xi_1$, we have

$$f^{(n)} = f_w e^{-\int An^{(n-1)}d\tau} + \int A f_0^{(n-1)} e^{-\int An^{(n-1)}d\tau} f_1^{(n)} d\xi_1 d\tau,$$

and we obtain the equation, linear in respect to $f^{(n)}$.

If for the equation, described above, is built up the chain by Markov, which would be identical to the trajectory of a trial molecule, then the functionals $\bar{\Psi}$ which are looked for, would be equal to the mathematical expectation for the accidental quantity gambled with an initial density of probability

$$f_w e^{-\int An^{(n-1)}d\tau}$$

and with an intermediate density

$$A f_0^{(n+1)} e^{-\int An^{(n+1)}d\tau}.$$

## 3.2	The Possibilities of the Scheme of Splitting for the Solution of Kinetic Equations

Considered in the work by Khlopkov and Serov[71,74] are the possibilities of the scheme of direct statistical modeling on the basis of the approximate kinetic equation. So let us consider the model equation by Krook:

$$\frac{df}{dt} = \gamma(f_0 - f),$$ (3.2.1)

$$\gamma = \frac{2}{\sqrt{\pi}} - \frac{n}{Kn},$$ (3.2.2)

where γ is the frequency of collisions.

If one addresses oneself to Chap. 1, the first stage of a scheme of splitting is described in accordance with (3.2.1). Let us write down the iteration scheme for the process as a whole:

$$\frac{f^{n+1} - f^n}{\Delta t^*} = \gamma^n(f_0^n - f^n)$$

or

$$f^{n+1} = f^n + \gamma^n \Delta t(f_0^n - f^n),$$
$$f^{n+1} = f^n(1 - \gamma^n \Delta t) + \gamma^n \Delta t f_0^n.$$ (3.2.3)

As it is not difficult to note, from Eq. (3.2.3) follows that the number of molecules, of which the velocity does not change after collision (that is of the molecules with the same distribution function f^n), makes up the $(1 - \gamma^n \Delta t)$th part of unity, while the similar number of molecules with the changed velocity makes up the $(\gamma^n \Delta t)$th part. Thus, one is able to compute within each cell the quantity $\gamma^n \Delta t$ and, choosing the accidental number R_i uniformly distributed between 0 and 1 for each particle in cell, one will change the velocity of ith particle, if $\gamma^n \Delta t \geq R_i$, and will not change it, if $\gamma^n \Delta t < R_i$.

The methodics developed was tested on the most simple example of heat transfer between two parallels plates heated to the temperatures T_w and T_L, $T_w/T_L = 4$. This example is.excellently demonstrating both all the positive and negative features of the methodics, because there was a possibility to compare the present results with those obtained earlier, in the works by Bird and Perepukhov.[11,59] By the way, the last of these authors was the first to apply the Bird's methods to a model equation.

The parameters of scheme were chosen in the following way. The length of a free path within the computational area is varying, and the cell's size is chosen to be smaller than minimal length of a free path λ_{min}. In the work discussed here, this size is chosen to be equal to $\frac{\lambda_{min}}{3}$. The value of step in time Δt^* was chosen proceeding from the considerations that it should be large as compared with time of collisions's duration and small as compared with duration of a free path. By each collision within the cell where this collision occurred, the step in time is made equal to

$$\Delta t_c = \frac{2}{N_c} \left[\pi b_{\max}^2 (2K)^{\frac{2}{\nu-1}} n g^{\frac{\nu-5}{\nu-1}} \right]^{-1}, \qquad (3.2.4)$$

where N_c is a total number of molecules within the cell, $b_{\max}$ is the maximal value of a dimensionless parameter of collision, K is the constant in the expression for the potential of interaction, n is the numerical density within the cell, g is the relative velocity, ν is a coefficient, equal to five for Maxwellian molecules. Within each cell the collisions are going on up to the moment when there is $\sum_{i=1}^{m} \Delta t_c = \Delta t^*$, and when two molecules after a single collision do not collide anymore. It was shown that with such a presentation of the process and of the quantity Δt^* in the form of a sum of "durations of single collisions" the modeled frequency of collisions coincides with a similar frequency computed from the Boltzmann's equation with the absolute error proportional to $D/N_p^{1/2}$, where D is dispersion and N_p — the possible number of pairs in the cell considered.

The temporal process goes on up to such a moment when the parameters, which are of interest for us, or the distribution function within each cell, come to a steady state. The distribution functions of the reflected molecules were presented as

$$f_L = n_L \left(\frac{1}{2\pi R T_L} \right)^{3/2} \exp \left(-\frac{\xi^2}{2 R T_L} \right), \qquad (3.2.5)$$

$$f_w = n_w \left(\frac{1}{2\pi R T_w} \right)^{3/2} \exp \left(-\frac{\xi^2}{2 R T_w} \right). \qquad (3.2.6)$$

The following notations are introduced here:

$$\begin{cases} T^* = \sqrt{T_L T_w}; \quad n_\infty = \dfrac{n_L + n_w}{2}, \\[2mm] K_n = \dfrac{\lambda_\infty}{L} = \dfrac{2}{\sqrt{\pi}} \dfrac{\sqrt{2 R T^*}}{n_\infty \sigma_0 L}; \quad \dfrac{T_w}{T_L} = R_T. \end{cases} \qquad (3.2.7)$$

By the interpretation of the equation by Krook it was assumed that after each collision the molecules acquire the locally Maxwellian distribution function, which means that after collision the molecules might be gambled in proportion to the locally Maxwellian distribution function. As it is known for the model equation, even after a single collision the system with an arbitrary distribution function comes to a nearly equilibrium state. Thus, one is ridded of the necessity to memorize the molecular distribution over velocities in each approximations, because after the collision the density of probability for velocity at the point of collision is equal to f_0.

The values of parameters n and T were chosen in accordance to the previous step in time Δt^* within the present cell. By such an approach the conservation laws are fullfilled in the cell only in aggregate, but not for each collision, and this situation is just real for the equation by Krook. The initial distribution was chosen in the form

$$f = n_\infty \left(\frac{1}{2\pi RT^*}\right)^{3/2} \exp\left(-\frac{\xi^2}{2RT^*}\right), \tag{3.2.8}$$

where ξ is dimensionless velocity related to $\sqrt{2RT^*}$. One gambles the initial velocities of $\sum_{i=1}^{m} S_i$ molecules (S_i — the number of molecules within ith cell, m — the number of cells) located between the plates. The step in time is chosen in the form

$$\Delta t^* = \frac{\alpha}{S}\sqrt{\pi}Kn\frac{L}{\sqrt{2RT^*}},$$

where S is a mean number of particles in cells, α is a factor which in different versions of computations varies from 1 to 4; the temporal duration of collision is chosen as

$$\Delta t_c = \frac{\sqrt{\pi}}{S_{it}^2}KnS\frac{L}{\sqrt{2RT^*}}, \tag{3.2.9}$$

where S_{it} symbolized the number of molecules in ith zone at the moment of time t. Within each cell is realized the proper number of collisions, and the difference $\Delta t_{it} = \Delta t^* - \sum_{i=1}^{k} \Delta t_{c_i}$ is memorized and is taken into account during the next collisional procedure.

As a result of computation, the following parameters are determined:

$$n = \frac{S_{it}}{S}n_\infty, \quad T = T^*\frac{2}{3}\sum_{S_{it}}\frac{V^2}{S_{it}},$$

$$Q = \frac{m}{2}n_\infty(2RT^*)^{3/2}\sum_{S_{it}}\frac{V^2V_x}{S}, \tag{3.2.10}$$

where n, T, and Q are, correspondingly, density, temperature, and heat flow.

After the moment when this process is brought to the steady state, one might reduce the error to the, practically, any desirable value, which is done by way of the averaging of all the parameters over the larger and larger number of the time intervals ΔT^*. The computations were conducted with the following values of the problem's parameters: $R_T = 4$, $S = 10$, $m = 10$, $Kn = 1$, and $S = 20$, $m = 20$, $Kn = 0.5$ (Figs. 3.2 and 3.3).

Now it is time to illustrate the possibilities of the scheme proposed. The parameters of this scheme were chosen in the following way: the cell's sizes were chosen proceeding from the condition that the distribution function f might be changed

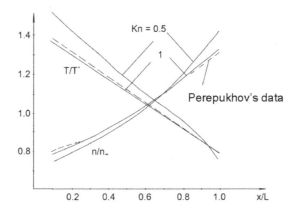

Fig. 3.2 Distribution of the density and of the temperature for $Kn = 0.5$ and 1, in comparison with the work by Perepukhov.

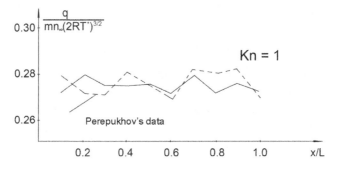

Fig. 3.3 Heat flow between the plates.

by the order at the length of the free path λ_∞. Therefore, the quantity Δx should be much less than λ_∞,

$$\Delta x \ll \lambda_\infty \Rightarrow \Delta x \ll KnL; \quad \Delta x = 0.1Kn. \tag{3.2.11}$$

The step in time was chosen from the condition that the cell's length Δx is passed over with a mean thermal velocity (stability condition),

$$\Delta t \le \frac{\Delta x}{\sqrt{2RT^*}}; \quad \Delta t \le \frac{\alpha Kn 0.1}{\sqrt{2RT^*}}. \tag{3.2.12}$$

Thus, the physical conditions of the problem and the mathematical properties of the equation lead to the automatical deterministical choice of the parameters of computational scheme. As results of the computations were determined the quantities n, T, and Q, calculated with the help of formulae (3.2.10). The computations were conducted with the following parameters of the problem:

$$Kn = 1, \quad R_T = 4, \quad S = 10, 8, 6, 4, 2, \quad m = 10;$$

$$Kn = 0.5, \quad R_T = 4, \quad S = 20.$$

The application of a new scheme of collisions permitted to diminish the number of particles within the cell for $Kn = 1$ from 10 down to 2. The further decrease of the number of particles in cell led to a large error.

Proposed for the decrease of the number of particles in cell were the following changes in the program:

(1) The temperature for gambling with the particle's velocity after the collision was brought not from the last temporal step Δt^*, but obtained from the averaging over all the steps Δt^* within each cell.

(2) The density of particles within the cell for the computation of the frequency of collisions with the subsequent modeling of collisional process was also obtained from the averaging over all the steps Δt^* within each cell.

The changes listed above permitted to drive the number of particles in cell to unity and to obtain the satisfactory results (Fig. 3.4).

The computations of the type described might be considered as a peculiar confirmation of the ergodical property — the fundamental law of the nature — at one hand, and of the common quality of the methods by Bird and by Haviland, at the other hand. And this consideration, in its turn, carries over the groundless of

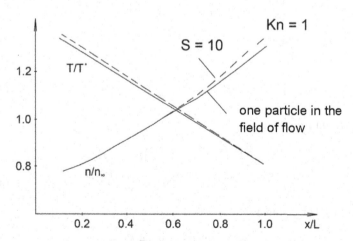

Fig. 3.4 Distribution of the density n and the temperature T for the cases of 10 particles in cell and one particle.

the methods of "trial particles" to the methods of modeling with the help of "the ensemble of particles".

The error of a computational scheme might be distinguished as deterministical one and statistical one. Deterministical error depends on the choice of the temporal step Δt^* and the size of the cell $\Delta x \cdot (\delta_1 \approx 0(\Delta t^*); \delta_2 \approx 0(\Delta x))$. The statistical error was calculated for $S = 8$, $Kn = 1$, $m = 10$. For the profile of n and $T_{CT} = 0.02$ (at the segment $N = 600, 800$), N — the number of temporal steps, $D = (0.02)^2 70019 \approx 0.03 \left(\delta_{CT} = \frac{3\sqrt{D}}{\sqrt{N}} \right)$. For the profile of $Q\delta_{CT} = 0.025$ (at the segment $N = 600, 80$), $D = (0.025)700/9 \approx 0.045$.

3.3 Increase of the Method's Rate of Convergence

The next stage of the present work was directed at the decrease of duration of the getting to the regime of a steady process. This is going to be done through the increase of exactitude of the approximation at each step Δt^*. Once again, let us turn to the model equation

$$\frac{df}{dt} = \gamma(f_0 - f) = J. \tag{3.3.1}$$

Using the scheme by Euler with repeated counting, one builds up the iterational process for Eq. (3.3.1) in the following way:

$$\begin{cases} f^{n+1/2} = f^n + J\dfrac{\Delta t}{2}, \\ f^{n+1} = f^n + J^{n+1/2}\Delta t. \end{cases} \tag{3.3.2), (3.3.3)}$$

From (3.3.2) one obtains

$$f^{n+1/2} = f^n \left(1 - \gamma^n \frac{\Delta t}{2}\right) + \gamma^n \frac{\Delta t}{2} f_0^n, \tag{3.3.4}$$

while the result from (3.3.3) is

$$\begin{aligned} f^{n+1} &= f^n + \left(f_0^{n+1/2} - f^{n-1/2}\right)\gamma^{n+1/2}\Delta t \\ &= f^n - f^{n+1/2}\gamma^{n+1/2}\Delta t + f_0^{n+1/2}\gamma^{n+1/2}\Delta t, \end{aligned} \tag{3.3.5}$$

and from (3.3.4) we obtain

$$f^n = \frac{f^{n+1/2} - \gamma^n \frac{\Delta t}{2} f_0^n}{1 - \gamma^n \frac{\Delta t}{2}},$$

while, substituting into (3.3.5), we have

$$
f^{n+1} = \frac{f^{n+1/2} - \gamma^n \frac{\Delta t}{2} f_0^n}{1 - \gamma^n \frac{\Delta t}{2}} - f^{n+1/2}\gamma^{n+1/2}\Delta t + f_0^{n+1/2}\gamma^{n+1/2}\Delta t
$$

$$
= f^{n+1/2} - \gamma^n \frac{\Delta t}{2} f_0 - f^{n+1/2}\gamma^{n+1/2}\Delta t \left(1 - \gamma^n \frac{\Delta t}{2}\right)
$$

$$
+ \frac{f_0^{n+1/2}\gamma^{n+1/2}\Delta t \left(1 - \gamma^n \frac{\Delta t}{2}\right)}{1 - \gamma^n \frac{\Delta t}{2}}.
$$

Neglecting of the terms of the small order $0(\Delta t^2)$ and making the expansion into series

$$
\frac{1}{1 - \gamma^n \frac{\Delta t}{2}} = \left(1 - \gamma^n \frac{\Delta t}{2}\right)^{-1} = 1 - \gamma^n \frac{\Delta t}{2} + 0(\Delta t^2),
$$

we obtain

$$
f^{n+1} = f^{n+1/2}[1 - (\gamma^{n+1/2} - \gamma^n/2)\Delta t]
$$
$$
+ (\gamma^{n+1/2} - \gamma^n/2)f_0^{n+1/2}\Delta t. \tag{3.3.6}
$$

Thus, the process might be modeled in such a way. One makes the flight of duration $\frac{\Delta t^*}{2}$, memorizes the frequency of collisions $v_j^{n+1/2}$ and realizes the collision of particles correspondingly to the previous scheme. After that one realizes the flight of duration Δt^* and, correspondingly to (3.3.6), one gambles the process of collisions. Once again, one chooses the accidental number R_i (which is uniformly distributed within the interval between 0 and 1) and changes the velocity of the ith particle within the cell, if

$$
\left(v_j^{n+1/2} - v^n/2\right)\Delta t \geq R_i
$$

or does not change it, if

$$
\left(v_j^{n+1/i} - v_j^{n+1}/2\right)\Delta t < R_i,
$$

where v_j^{n+1} is the frequency of collisions within jth cell, at the $(n + 1)$th step in time Δt^*. This method was approbated by the solution of the problem of a heat transfer between the plates. The scheme's parameters were chosen corresponding to the parameters of a preceding problem. The iterational process is, in its essence, excluding the process of splitting in time Δt^* of the motion of particles and their collision. Thus, we have not any limitation for Δt^* being over the stability criterium. In the present work, the quantity Δt^* was chosen in the form

$$
\Delta t^* = \frac{\alpha Kn}{\sqrt{3RT^*}}. \tag{3.3.7}
$$

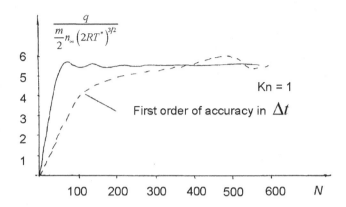

Fig. 3.5   Dependence of the solution's convergence on the number of temporal steps.

The factor α varied between 0.5 and 1. In Fig. 3.5, computational results concerning the dependence of the convergence of the present scheme on the number of temporal steps in comparison with traditional scheme (for $Kn = 1$, $S = 8$, $m = 10$), indicate the acceleration of that convergence, approximately, in one order.

3.4   Method by Belotserkovskii and Yanitskii

The probabilistical nature of the dynamics of rarefied gases, which is so important for application and development of the numerical schemes Monte Carlo, follows quite naturally from the general principles of kinetic theory and statistical physics. The considerations given below might be looked at, also, as the levels of carefulness of the description of the dynamics of a large molecular system, and in the later presentation we shall use these levels for the building-up of the effective methods of statistical modeling.

As it was noted in the introduction, the most detailed level of the description contains in a dynamical system. For the description of such a system, which consists of a large number of elements N (the molecular gas represents in itself just such a system with $N \approx 10^{23}$ molecules) it is necessary to prescribe the initial coordinates and velocities of each molecule $(\vec{r}_j, \vec{v}_j)$ and the equations describing the evolution of that system:

$$m\frac{d^2\vec{r}_j}{dt^2} = \sum_{i \neq j}^{N} R_{ij}. \tag{3.4.1}$$

Solution of such a system of equations presents in itself quite unrealistical problem, even for the extremely rarefied gas: at the height of 400 km (the most popular orbits of satellites) one cubic centimeter contains 10^9 molecules. For that reason one turns oneself to the less complete — statistical — description of the system's behavior.

Following the formalism by Gibbs one considers not a single system, but the ensemble of systems in $6N$-dimensional Γ-space distributed according to the N-particle distribution function $F(t, \vec{r}_1, \vec{r}_2, \ldots, \vec{r}_N, \vec{v}_1, \vec{v}_2, \ldots, \vec{v}_N) = F_N$, of which the sense coincides with the probability of this system being at the moment of time t at the point $\vec{r}_1, \vec{r}_2, \ldots, \vec{r}_N, \vec{v}_1, \vec{v}_2, \ldots, \vec{v}_N$ in vicinity of $d\vec{r}_1 \cdots d\vec{r}_N d\vec{v}_1 \cdots d\vec{v}_N$:

$$dW = F_N d\vec{r}_1 \cdots d\vec{r}_N d\vec{v}_1 \cdots d\vec{v}_N.$$

Such an ensemble is described by the well-known equation by Liouville:

$$\frac{\partial F_N}{\partial t} + \sum_{i=1}^{N} v_i \frac{\partial F_N}{\partial r_i} + \sum_{i \neq j}^{N} \sum_{i=1}^{N} \frac{R_{ij}}{m} \frac{\partial F_N}{\partial v_i} = 0. \tag{3.4.2}$$

From that point on the Liouville equation and all the other kinetic equations following from the Bogoljubov's chain, including its last link — Boltzmann's equation, possess the probabilistical nature. And in spite of the fact that Eq. (3.4.2) is simpler than the system (3.4.1), it takes into account the N-particle collision of molecules and is also extremely complicated for the practical analysis. The turning to the less detailed level of description is connected with the further coarsening of the system's description with the help of the s-particle distribution functions $F_s = \int F_N d\vec{r}_{s+1} \cdots d\vec{r}_N d\vec{v}_{s+1} \cdots d\vec{v}_N$, which determine the probability of a simultaneous revelation of s particles, independently of the state of other N-s particles. Following the Bogoliubov's ideas, one obtains the chain of interconnected equations:

$$\frac{\partial F_s}{\partial t} + \sum_{i=1}^{s} v_i \frac{\partial F_s}{\partial r_i} + \sum_{i=1}^{s} \sum_{j \neq i}^{s} \frac{R_{ij}}{m} \frac{\partial F_s}{\partial v_i}$$

$$= -\sum_{i=1}^{s} (N - s) \frac{\partial}{\partial v_i} \int \frac{R_{i,s+1}}{m} F_{s+1} dr_{s+1} dv_{s+1}, \tag{3.4.3}$$

which continues up to the one-particle distribution function $F_1 = f(t, \vec{r}, \vec{\xi})$ of the Boltzmann's gas taking into account the binary collisions only,

$$\frac{\partial f}{\partial t} + \vec{\xi} \frac{\partial f}{\partial \vec{r}} + \frac{R_{12}}{m} \frac{\partial f}{\partial \vec{\xi}} = -\frac{\partial}{\partial \vec{\xi}} \int \frac{R_{12}}{m} F_2 d\vec{r}_1 d\vec{\xi}_1.$$

Following the Boltzmann's reasoning, let us assume that the molecules are spherically symmetric, and accepting the hypothesis of a molecular chaos, $F_2(t, \vec{r}, \vec{v}_1, \vec{v}_2) = F_1(t, \vec{r}, \vec{v}_1) F_1(t, \vec{r}, \vec{v}_2)$, we shall come to Eq. (3.4.2).

Seems to be rather interesting the particular case of a Liouville equation (3.4.2) and a Bogoljubov's chain (3.4.3) for the spatially uniform consideration of the gas consisting of the limited number of particles, which at the final link, corresponding to the two-particle collisions, leads to the well-known equation by Kac-"Master Equation"[7]:

$$\frac{\partial \phi_1(t, \vec{\xi}_1)}{\partial t} = \frac{N-1}{N} \int [\phi_2(t, \vec{\xi}_1', \vec{\xi}_2') - \phi_2(t, \vec{\xi}_1, \vec{\xi}_2)] \cdot g_{12} d\sigma_{12} d\vec{\xi}_2, \qquad (3.4.4)$$

where φ_1 and φ_2 are one-particle and two-particle distribution functions.

Unlike the Boltzmann equation, Eq. (3.4.4) is linear one, and this fact will be used by the creation and justification of the effective numerical schemes of direct statistical modeling.

Up to the present time, the unconditional priority in the rarefied gas dynamics belongs to the method by Bird, the modified version of which made by the Russian scientists (see, for example, Refs. 10–21) permitted to increase just in several orders the effectiveness of that method.

The essence of Bird's method is found in that the system's evolution at some small interval of time Δt is splitted in two clearly comprehended processes: (1) the relaxation in accordance with a collisional operator in the kinetic equation,

$$\frac{\partial f}{\partial t} = J(f)$$

and (2) the free-molecular transfer

$$\frac{\partial f}{\partial t} = -\vec{\xi} \nabla f.$$

This is the well-known scheme of the first-order splitting over Δt for the arbitrary operator equation, but in the present case it is quite tempting due to the fact that it is splitting the dynamics of extremely complicated kinetical system into two clearly understandable physical processes. The distribution function is modeled by N particles, which at the first stage are colliding between them within each cell in accordance with the frequency of collisions, during the time interval Δt, while at the second stage they fly during Δt over the distance $\vec{\xi}_j \Delta t$.

The central point of the method of nonstationary statistical modeling appears to be the procedure of the computation of collisions. The pair of particles is chosen for collision accordingly to the frequency of molecular collisions and independently of the distance between the molecules colliding within the present cell. The velocities of particles after the collision are chosen in accordance with laws of interaction of the molecules. And in spite of the fact that the effectiveness of the method depends on rather numerous amount of the parameters of computational scheme (steadification, temporal splitting, coming to the stationary regime, step in time, form of the spatial network, and so on), the most important works on the method's improvement are devoted to the perfection of a collisional procedure and to the decrease of statistical error of the scheme. These factors permit to diminish the number of particles in cells and, correspondingly, to diminish both the necessary operative memory of the computer and the computational time. Thus, proposed in Ref. 18 was the modification of the collisional procedure for one particular case of the Maxwellian molecules, and with this modification the results of computations are, practically, independent of the number of particles in cell when it is varies from 40 to 6 (by the everyday computations the number of particles in cell is of the order of 30). Proposed in Refs. 11–16 is the general method, independent of the type of molecules, by which at the stage of collisions the subsystem of particles within each cell is considered as N-particle model by Kac, Eq. (3.4.4).

The modeling of collision is reduced to a statistical realization of evolution not of the Boltzmann equation, but of the model (3.4.4), during the interval Δt. The time of collisional duration in Kac's model is computed accordingly to the statistics of collision in the perfect gas with the help of a scheme by Bernoulli. This scheme permits to use the essentially diminished number of particles in cell and the smaller step of the computational network. As it was shown by the analysis, the results of computations are, practically, independent of the number of particles in cell, down to the value of 2. As a matter of fact, the Boltzmann equation demands with necessity the realization of the hypothesis of molecular chaos, while at the number of particles in cell which might be reached on the contemporary computers, this hypothesis might be realized with a systematic error, and as concerns Eq. (3.4.4), it is free of such a demand and, therefore, the stage of collision is computed as Markovian process. At the other hand, by $N \to \infty$ there exists the equivalence between Kac's model and spatially uniform Boltzmann equation.

Thus the approach developed by Belotserkovskii and Yanitskii creates, first, the method of construction of the effective numerical schemes which would provide the possibility of solution of the three-dimensional problems of aerodynamical flow about bodies, and, second, solves the most important methodological problem of the equivalence of the numerical method to a solution of the kinetic equation.

Chapter 4

Modeling of the Flow of Continuous Media

4.1 Procedure of the Monte Carlo Methods for Modeling the Flows of Rarefied Gas and Continuous Medium

The study of problems of the rarefied gas dynamics is, as a rule, connected with a solution of kinetic equations for the distribution function. The widespread application of Monte Carlo methods in that area is conditioned, on one hand, by the complicated multi-dimensional structure of the kinetic equations and, on the other hand, by the abundance of information contained in the distribution function. The statistical description of the state of a gas permits to spread the Monte Carlo method into the area of continuous medium, where such a way of problem's solution is nontraditional and scarcely justified, but, nevertheless, permits to develop the general approach to the modeling of different regimes of flow.

Considered here is a statistical procedure of the modeling of gaseous flows by the chosen in certain way set of particles, the totality of which determines the state of a medium. The field of a flow is divided into cells, the particles are located in this field and endowed by a number of properties as, for example, by a mass m, velocity ξ, and coordinate x. For the rarefied gas, the knowledge of these particles is sufficient if one wishes with the proper number of particles N to characterize the distribution function and through the averaging over the particles to determine the macroparameters. In the area of flows, subject to description by the moments of distribution function, the particles, which determine the state of a gas, possess the somewhat different set of control properties ψ, as it is done, for example, in the method for Euler's equations. In this case the velocities of particles within the cell correspond to the velocity of flow and, moreover, the particles possess the internal energy ε. By the analogy with kinetical distribution function, one might consider the distribution function to be in the form

$$f \approx \prod_{i=1}^{N} \delta(\xi_i - u)\delta(\varepsilon_i - E),$$

where u and E are the velocity and internal energy of a gas within the cell. Then the determination of macroparameters is realized by a standard way:

$$\Psi = (f, \psi) \approx \frac{1}{N} \sum \psi_i.$$

In the gaseous flows described by the equations of moments of the higher order, the particles might possess the larger set of the internal control properties $P_{ij}, q_i, \ldots$.

The trajectories of particles are described, generally speaking, by the differential equations of motion

$$m\frac{d\vec{\xi}_i}{dt} = \sum_j F_{ij}, \quad \frac{d\vec{x}_i}{dt} = \vec{\xi}_i, \tag{4.1.1}$$

where F_{ij} is a force acting on the particle j from the part of particles i. The evolution of the temporal history of particles proceeds into the process of succession of the finite intervals of duration Δt, during which are measured the velocities and internal control properties of the particles, and after that is realized the transfer of these properties along the trajectories.

The term corresponding to the force in Eq. (4.1.1) symbolizes in the rarefied gas the inter-pair interaction of the particles with the corresponding frequency of collision ν. Thus for model kinetic equation, for which in the preceding chapter was developed the proper procedure, one has

$$\frac{df}{dt} = \nu(f_0 - f). \tag{4.1.2}$$

By the increase of the temporal step by the quantity Δt the trial particle realizes with a probability, proportional to $(1 - \nu\Delta t)$, the free flight along the trajectory,

$$x_j^{n+1} = x_j^{n+1} + \xi_j^{n+1}\Delta t, \tag{4.1.3}$$

but otherwise its velocity is changed in accordance with the equilibrium distribution function f_0.

In the case of equations of moments the term with force in (4.1.1) is determined by the gradients of macroparameters. The trajectories of particles are built up according to a scheme:

$$\xi_j^{n+1} = \xi_j^{n1} + \phi(\vec{\xi})\Delta t, \quad \vec{x}_j^{n+1} = x_j^n + \left(\xi_j^n + \xi_j^{n+1}\right)\frac{\Delta t}{2}, \tag{4.1.4}$$

while the variation of internal control properties ε is also conditioned by the terms with gradients in the corresponding equations:

$$\varepsilon_j^{n+1} = \varepsilon_j^n + \phi''(\vec{\xi})\Delta t.$$

In the case of a statistical modeling, which is considered here, the error $\delta\bar{\psi}$ for the numerical functions within the cell will be equal to

$$\delta\psi \approx O\left(\frac{1}{\sqrt{N}}\right).$$

By the passing of particles from one cell to another the order of error should be conserved.

Let us assume that the order of error by the computation should not be higher than Δt. In spite of the fact that the accuracy of the problem's solution is, eventually, determined by the accuracy of computation for the macroparameters $\delta\bar{\psi}$, there exist a certain interference between the sources of statistical error and deterministical one. Therefore, it is necessary that these errors are of one and the same order:

$$O(\Delta t) \approx O(\Delta x) \approx O\left(\frac{1}{\sqrt{N}}\right), \qquad (4.1.5)$$

that is, the discrete trajectory of the particle should not differ from the real one at the characteristical dimension of flow larger than Δx. Just by this circumstance is explained the use of the scheme of the second order (4.1.4) in the case of continuous medium. Only in such a case we have

$$\delta x^n = O(n^2 \Delta t^2 + (\delta\varphi + \Delta t)).$$

From the considerations listed above follows the additional relation between the number of temporal steps and the length of a step:

$$n\Delta t \approx O(1). \qquad (4.1.6)$$

These conditions should be supplemented by the condition of stability:

$$|\xi_j|\frac{\Delta t}{\Delta x} < 1,$$

which in application to the statistical methods is frequently changed for a softer condition:

$$F = \left\{|\xi_j|\frac{\Delta t}{\Delta x} > 1\right\} \ll 1. \qquad (4.1.7)$$

Since the change of the velocities of particles in rarefied gas occurs instantaneously, in the relation (4.1.3) it is possible to take the scheme of the first order of accuracy.

Thus, the conditions (4.1.5)–(4.1.7) and the scheme as (4.1.3) or (4.1.4) might serve to the determination of a computational procedure. The procedure described is widely used for the solution of problems in the dynamics of rarefied gases.

The possibilities of application of the method to the case of a continuous medium were checked through the solution of a problem of the flow between two parallel infinite plates. In hydrodynamical case, the one-dimensional equations are described in the form

$$\frac{\delta u}{\delta t} = \frac{1}{\rho}\frac{\delta}{\delta y}\mu\frac{\delta u}{\delta y}. \tag{4.1.8}$$

Modeling of the flow might be presented in such a way. The field of flow is divided in cells, within each of which the distribution function of the particles has a form

$$f \approx \prod_{i=1}^{N} \delta(\xi_i - u).$$

Thus, the particles with one and the same mass possess the internal energy and velocity. The variation of control properties is realized in accordance to Eq. (4.1.8).

Having the estimate of the order of accuracy, one makes the division of the field of flow in several dozens of cells and conducts the tests, of which the number is of the order of ten thousands.

By the solution of the problem of a flow by Couette there is no passing of particles from one cell to another, and the relaxation is realized in accordance to Eq. (4.1.8). The results of solution of the problem by Couette are presented in Fig. 4.1.

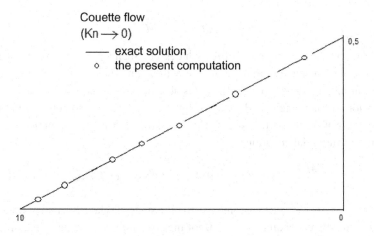

Fig. 4.1

4.2 Method "Relaxation–Transfer" for a Solution of the Problems of Gas Dynamics in the Wide Range of the Degree of Rarefaction of a Medium (see Kogan *et al.*[83])

By the solution of the problems of gas dynamics, which are described by the kinetic equation

$$\frac{\delta f}{\delta t} = -\xi \nabla f + J(f), \qquad (4.2.1)$$

one usually divides the process of solution into two independent stages within the small interval.

The first stage–stage of transfer:

$$\frac{\delta f}{\delta t} = -\xi \nabla f. \qquad (4.2.2)$$

The second stage–stage of relaxation:

$$\frac{\delta f}{\delta t} = J(f). \qquad (4.2.3)$$

Both these stages possess the clear physical sense, and it is, actually, the collisionless transfer of particles and the spatially uniform relaxation. Such an approach is successfully used, for example, in the method of a direct statistical modeling.

In the present book, one uses the scheme of solution for the momental presentation of the kinetic equation. In this case is considerably simplified the most complicated part of the solution of kinetical problems — computation of the integral of collisions. The distribution function is considered in some approximational form:

$$f = F(\xi, M^{(0)}, \ldots, M^{(k)}), \qquad (4.2.4)$$

where $M^{(k)}$ is a kinetic moment of the kth degree. Such a presentation provides a whole set of conveniences in the concrete realization. Thus, being at the stage of relaxation in accordance to (4.2.3), one is able to pass to the macro-level of the description of phenomenon. By way of limiting oneself by the consideration of the moments of the third order one might obtain for the Maxwellian molecules the system of differential equations:

$$\frac{\delta M^{(k)}}{\delta t} = \alpha M^{(k)}, \quad M^{(j+1)} = M^{(j+1/2)}(1 + \alpha \Delta t), \qquad (4.2.5)$$

where M^j and $M^{j+1/2}$ are the moments after the stage of relaxation and of transfer, correspondingly, the coefficient $\alpha = 0$ for the equations containing concentration n, mean velocity u, and temperature T within the cell, while $\alpha = -1/\tau_P$ for the elements of stress tensor P_{ij} and $\alpha = -2/(3\tau_P)$ for the components of the flow of heat q_i, where τ_P is the duration of the time of relaxation.

The stage of transfer might be realized at the level of the approximational distribution function with the moments taken from the preceding step.

The choice of distribution function (4.2.4) depends on the physical peculiarities of the problem considered. Thus, for example, by the study of strongly rarefied flows, where the determination factors are the boundary conditions, it would be convenient to choose (4.2.4) in the form of a function with directional ruptures. In the flows of nearly continuous medium it would be more convenient to take the continuous form of a distribution function, presenting it as the expansion over the small parameter.

The computational scheme is built up in the following way. The field of flow is determined and at its boundaries is prescribed the form of a distribution function, while in the internal part of flow one chooses the approximational function of distribution (4.2.4). The field of flow is divided into the cells of the size Δx. One chooses the temporal interval $\Delta \tau$, within duration of which the process is splitted into two stages: the free flight (4.2.2) and the uniform relaxation (4.2.3).

The error of the method δ is determined by the degree of success in the choice of the approximational distribution function (4.2.4). As for the type of the numerical scheme, it appears to be the first-order scheme in respect of accuracy, $\delta = o(\Delta t, \Delta x)$, and it is necessary to take into account by the approximation of spatial derivatives by the future differences and by the observation of the stability condition $|\xi| \cdot \Delta t / \Delta \chi \leq 1$.

Depending on the degree of complication and on the sizes of a problem, the stage of transfer (4.2.2) might be realized either at the level of macroparameters, or at the level of distribution function. In one-dimensional problems, the stage of transfer would be convenient to realize at the moment's level, by way of integration of (4.2.2) with the proper weights. For the two- or three-dimensional problems would appear essentially more important both the complication and the nonlinearity of the momental equations of transfer, and for this reason in these cases would appear as effective to build up the direct solution of a linear equation (4.2.2). This might be done, for example, by way of the solution in each cell of the system of linear finite-differential equations for a set of discrete ordinates, or by way of modeling of the process of transfer using Monte Carlo method.

At the stage of relaxation, when approximating the distribution function (4.2.4) by the continuous function, as it is done, for example, in the classical method by Grad, one will immediately obtain the values of moments (4.2.5). In all the other cases, as, for example, using the approximation by Lees, one meets the necessity of inversion of the system of algebraical equations.

The approbation of method will be done on the problem of the flow and heat transfer between the infinite parallel plates. There is a usual formation of this problem. The coordinates origin is chosen to be on one of the plates. The x-axis is perpendicular to the plates. The velocities of plates and their temperatures are, correspondingly, $u_{W1,2}$, $T_{W1,2}$.

The approximational form of the distribution function is chosen in two versions. In the form of the two-flow Maxwellian distribution or, so-called, approximation by Lees, and in thirteen-moment approximation by Grad, the nondimensionalization is realized over the parameters at one of the boundaries, and then one obtains the Boltzmann equation for distribution function:

$$\text{by Lees} \; - f^{\pm} = \frac{n^{\pm}}{(\pi T^{\pm})^{3/2}} \exp\left(-\frac{c^2}{T^{\pm}}\right), \qquad (4.2.6)$$

$$\text{by Grad} \; - f = \frac{n}{(\pi T)^{3/2}} \exp\left(-\frac{c^2}{T}\right)\left\{1 + \frac{P_{ij}}{PT}C_iC_j - \frac{4q_i}{PT}\left(1 - \frac{2}{3}\frac{C^2}{T}\right)G\right\}. \qquad (4.2.7)$$

The problem is one dimensional, and by this reason the transfer is easily realized on the level of macroparameters.

The stage of a free flight (4.2.2) for approximation by Lees (4.2.6):

$$(n^{\pm})_i^{j+1/2} - (n^{\pm})_i^j = \pm\frac{\Delta t}{\Delta x}\cdot\frac{1}{\sqrt{\pi}}((n^{\pm}\sqrt{T^{\pm}})_{i\pm 1}^j - (n^{\pm}\sqrt{T^{\pm}})_i^j),$$

$$(n^{\pm}\sqrt{T^{\pm}})_i^{j+1/2} - (n^{\pm}\sqrt{T^{\pm}})_i^j = \pm\frac{\Delta t}{\Delta x}\cdot\frac{5\sqrt{\pi}}{4}((n^{\pm}(T^{\pm})^2)_{i\pm 1}^j - (n^{\pm}(T^{\pm})^2)_i^j).$$

For the approximation by Grad the corresponding equations of transfer are obtained by means of integration of the equation in differences,

$$\frac{f_i^{j+1/2} - f_i^j}{\Delta t} = |\xi_x^{\pm}|\frac{f_{i\pm 1}^j - f_i^j}{\Delta x}$$

with the proper weights $\varphi(\xi)$ for the neighboring cells, where

$$\phi(\xi) \in \left\{1, \xi_x, \xi_x^2, \xi_y, \xi_x\xi_y, \xi^2, \ldots\right\},$$

$$\int_{-\infty}^{\infty} f_i^{j+1/2}\phi d\xi = \int_{-\infty}^{\infty} f_i^{j+1/2}\phi d\xi + \left(\int_{-\infty}^{\infty} f_{i-1}^j \xi^+ \phi\, d\xi + \int_{-\infty}^{\infty} f_{i-1}^j \xi\vec{v}\phi d\xi\right.$$

$$\left. + \int_{-\infty}^{\infty} f_i^j |\xi^{\pm}|\phi d\xi\right)\frac{\Delta t}{\Delta x}.$$

On the basis of results of a transfer one determines the data for the stage of relaxation:

$$n_i^{j+1} = n_i^{j+1/2}, \quad u_i^{j+1} = u_i^{j+1/2}, \quad p_i^{j+1} = p_i^{j+1/2},$$

$$p_{xy_i}^{j+1} = p_{xy_i}^{j+1/2}\left(1 - \frac{2}{3}\frac{\sqrt{\pi}}{Kn}n_i^{j+1/2}\Delta t\right),$$

(4.2.8)

$$q_{x_i}^{j+1} = q_{x_i}^{j+1/2}\left(1 - \frac{2}{3}\frac{\sqrt{\pi}}{Kn}n_i^{j+1/2}\Delta t\right),$$

and one obtains the possibility to pass to the stage of transfer.

To restore the two-flow distribution function, it is necessary at the $(j+1)$th step to solve the system of algebraic equations containing $n^\pm$, $u^\pm$, $T^\pm$ within each cell. For example, in the case of a heat transfer one obtains the system

$$n = \frac{1}{2}(n^+ + n^-),$$

$$nT = \frac{1}{2}(n^+ T^+ + n^- T^-),$$

$$q_x = \frac{1}{2\sqrt{\pi}}(n^+ (T^+)^{3/2} + n^- (T^-)^{3/2}),$$

which is solved by the method of successive approximations.

Using the thirteen-moment approximation, it is possible to solve the gas-dynamical problems with the help of Monte Carlo method. Aiming at that, in accordance with distribution function are gambled the velocities of particles and their coordinates, and after that the particles are transferred during the time $\frac{\Delta t}{2}$: $x^{j+1/2} = \xi \cdot \frac{\Delta t}{2} + x^j$. The boundaries should be taken into account by the special way. Further on, knowing the data on velocities and coordinates of the particles after the stage of transfer, one calculates n, u, T, q_x, p_{ij} for each cell and realizes the relaxation in accordance with (4.2.8).

With the data obtained after the relaxation one begins for each cell to build up a new distribution function (4.2.7), and in accordance with it and gambling the velocities and coordinates, one realizes the transfer, and so on.

In Ref. 83 are given the examples of solution of the problem of heat transfer by the various Knudsen numbers, namely, $(\infty, 10, 1, 0.5, 0.1, 0.01$, thirteen-moment approximation, NS — approximation, nonviscous limit — $Kn = 0$) and by the various temperature relations $(T_{W2}/T_{W1} = 4, T_{W2}/T_{W1} = 16)$ the authors conducted the comparison with the results by Tcheremissin, Vlasov, Liu, and Lees.

And as it should be expected, for $Kn > 1$ there was obtained a good correlation with the results of paper by Liu and Lees for the two-flow approximated function (4.2.6). For that approximation, a good agreement was obtained with the exact numerical solution of the Boltzmann equation. For the smaller Knudsen numbers, the best agreements with the results of Ref. 83 provides the thirteen-moment approximation, which works quite badly at the large Kn. In particular, by the tending of Kn to the free-molecular limit the error in temperature is over 50%. The consideration of approximation with moments of fourth order leads to an insignificant diminishment of error. The explanation of that is connected with the fact that it is impossible to approximate by such a way the broken distribution function.

Finally, one should make a conclusion that using the methodics described and making the proper choice of the approximational function one would be able to solve, quite effectively, the problems of rarefied gas dynamics by the arbitrary numbers Kn. In particular, this method permit to model the continuous flows, as well, for which the form of a distribution function is known exactly (Maxwell, Navier–Stokes, . . .). Thus the cases of such an approach to the modeling of the continuous medium are described in the papers by Chetverushkin, Elizarova, Pavlov, Abalkin, Voltchinskaja, and Pullin.[81,107−110]

In the number of cases, the modeling of the equations of continuous medium at the level of a distribution function proves to be more convenient, because such an approach changes the complicated nonlinear system of the equations in partial derivatives by the simple linear equation corresponding to the flight of particles from one cell to another.

4.3   Modeling of the Flows of Nonviscous Perfect Gas

As it was already noted earlier, the stage of transfer might be conducted at the level of a distribution function. The medium to be modeled might be changed for a system of N particles, which are distributed at the initial moment of time within the cells of the immobile Eulerian network and with the velocities which are modeled with the help of the distribution function in this particular cell. Following the method described in Ref. 74 we are going to model the transfer of particles, of which the velocities are distributed in accordance to the Maxwellian distribution function. At the stage of transfer, the particles fly along the trajectories without collisions between them and interacting with the boundaries only. The collisions of particles occur during the time of relaxation. At this stage, the function of distribution of the particles acquires the Maxwellian form. Thus, by the way of forcing the form of a distribution function for the stage of transfer one might model the natural process. Considered below is the method's application to the one- and two-dimensional problems. Moreover, as an example of taking into account the various effects, one gives the estimation of the influence of the rotational degrees of freedom. It is assumed that the temperature corresponding to the rotational degrees of freedom is equal to that corresponding to the transitional degrees of freedom. This means that at the stage of relaxation is realized the complete interchange of energy between the different degrees of freedom. The energy corresponding to the rotational degrees of freedom is equal to $e = \frac{\gamma T}{4}$, where $\gamma = \frac{5-3\nu}{\nu-1}$ is the number of internal degrees of freedom.

The methods of statistical modeling possess both the advantage and convenience in their comparison with finite-difference methods, as concerns the formulation of the boundary conditions. The boundary conditions are set at the level of a distribution function. This leads to a possibility of the solution of problems of the flow about bodies of the complicated configuration with application of the network of rectilinear form. Moreover, the method described above possesses such an advantage that it does not demand to memorize the velocities and coordinates of all the particles. It would be just sufficient to memorize only the macroparameters in cells.

General formulation of the problem. The field of flow $\Omega(t)$ with a boundary $\partial\Omega(t)$ is divided in cells of the size $\Delta\vec{x}$. The time is considered as a discrete object:

$$t_j < t < t_{j+1}, \quad t_j = j\Delta t.$$

The choice of quantities Δt, $\Delta\vec{x}$ will be discussed by the solution of concrete problems.

1. Gambling of the velocities $\vec{\xi}_n$ and coordinates of the particles $\vec{x}_n (0 \le n \le N_i^j)$ in accordance with the macroparameters N_i^j, u_i^j, T_i^j and the Maxwellian distribution function.
2. Transfer of the particles, $\vec{x}_n^{j+1} = \vec{\xi}_n^j \Delta t + \vec{x}_n^j$. If the particle collides with the boundary $\Omega(t)$, then it is necessary to take into account such an interaction.
3. Taking into account, if necessarily, the particles entering the area $\Omega(t)$ and those going out of $\Omega(t)$.
4. The determination of the number of cell, into which the particle has flown.
5. Added to the macroparameters in the cell under consideration are the parameters of the particles which flew into that cell.
6.

$$U_i^{j+1/2} = \frac{1}{N} \sum_{n=1}^{N} \xi_{i,n}^{j+1/2}, \quad E_i^{j+1} = E_{c,i}^{j+1/2} + E_{t,i}^{j+1/2},$$

$$E_{\bullet c,i}^{j+1/2} = \frac{1}{2} \sum_{n=1}^{N} (\xi_{i,n}^j - U_i^{j+1/2})^2, \quad p_i^{j+1} = \frac{4}{3+\nu} \frac{E_i^{j+1}}{N_i^j},$$

$$E_{\mathrm{t}} = \sum \frac{\nu T_{ij}}{4}.$$

7. The computation of the averaged macroparameters and completion of the statistics for stationary problems.
8. Return to the point 1.

It is necessary to note that the algorithm described proves to be conservative in respect to mass, momentum, and energy under condition of the observance of conservativeness at the stage of gambling with velocities of particles. Here is used the algorithm which is conservative in respect to the mean velocity and energy.

Problem of the motion of a piston in nonviscous perfect gas. The problem is solved for the monatomic gas and for the diatomic one. As it is indicated in Ref. 81, memorized by the computations are the coordinates of all the particles, which fact is, naturally, limiting the area of applicability of the method. By the solution of problems with the help of method described in the present paragraph it would not be necessary to memorize coordinates of the particles. This property leads to a possibility of using this method by the solution of two- and three-dimensional problems, both stationary and nonstationary ones. The gambling of the particle's coordinates is conducted here in accordance with the first derivative of the concentration of particles, according to a function

$$f = \frac{n_{i+1} - n_{i-1}}{2(n_{i+1} + n_{i-1})}(x - 0, 5) + 1,$$

where $0 \le x \le 1$, x being the coordinate within the cell.

Considered here is the one dimensional, nonsteady problem of the forming of the jump and its motion before the piston, which moves with a constant speed. The problem is solved for the diatomic gas ($\nu = 2$) and for the monatomic one ($\nu = 0$). At the initial moment of time $t = 0$ the piston stays in the point $x = 0$ and instantaneously acquires the speed $u_p > 0$.

At the moment of time $t = 0$ gas is unperturbed within the area ($\vec{u}_i = 0$, $\vec{T}_i = 0$). Before the piston is formed the shock wave, which moves at the speed $u_s > u_p$. The area of perturbation is located between the piston and the shock wave, and therefore it might be possible to restrict the area of flow and guess that the perturbation does not reach the point $x = 1$. The problem possesses an exact solution:

$$U_p = \frac{\sqrt{\gamma}}{2}\left(M_s - \frac{\frac{\gamma-1}{2}M_s^2}{\sqrt{\frac{1}{\gamma M_s^2} - \frac{\gamma-1}{2}}}\right)\frac{\sqrt{2(\gamma-1)(M_s^2-1)(1+\gamma M_s^2)+1}}{(\gamma+1)^2 M_s^2},$$

$$\frac{n_1}{n_2} = \frac{M_s^2 - 1}{1 + \frac{\gamma-1}{2}M_s^2} + 1, \qquad \frac{T_1}{T_2} = \frac{2(\gamma-1)(M_s^2-1)(1+\gamma M_s^2)}{(\gamma+1)^2 M_s^2},$$

$$\frac{p_1}{p_2} = \frac{2\gamma(M_s^2-1)}{\gamma+1} + 1.$$

To guarantee the fulfillment of a nonflow-through condition $u_n = 0$ using the Monte Carlo method it would be necessary to consider the condition of the particles reflection from a piston and at $x = 1$ to be specular, that is

$$\xi_n^i = 2u_p - \xi_n (x = x_p); \quad \xi_n^i = -\xi_n, \quad (x = 1).$$

By the solution of the problem under consideration the area of a flow $x \in (0, 1)$ was divided in 40 cells. At the moment $t = 0$ within each cell were located 200 particles. The problem was solved for $\gamma \in \{\frac{5}{3}; 1; 4\}$ and for the numbers $M_s = \{2.3; 3.0; 7.0\}$. Dealing with the problems described by the Euler equation, the sizes of cells might be chosen arbitrarily. The choice of Δx depends on that condition, which should be the accuracy of a solution (along how many cells in the present problem will be "spread" the shock wave). The step in time is determined in such a way that one would have $p = \frac{\Delta t}{\Delta x} u_s \approx 1$.

The results are presented in Fig. 4.2. To guarantee the acceptable degree of accuracy it is sufficient to observe the condition $p \leq 1$.

The flow in the nozzle by Lavalle. In the case when the macroparameters experience just small variations along the cross section, where the velocity is directed, practically, along the nozzle, the problem might be reduced to the one-dimensional form (this fact, really, takes a place because the area of a cross section varies along

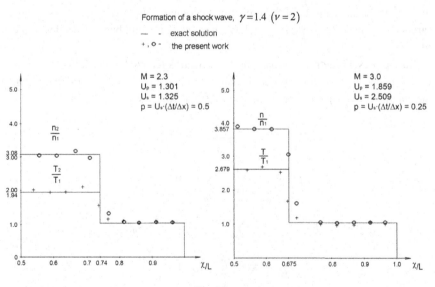

Fig. 4.2

the nozzle's length just slowly). The form of a nozzle was chosen according to the following expression: $S = 1 - (1 - S_{cr})\sin \pi x$, while S_{cr} is a critical cross section

$$S_{cr} = \frac{M\left(\frac{\gamma+1\Gamma}{2}\right)^{\frac{\gamma+1}{2(\gamma-1)}}}{\left(1 + \frac{\gamma+1\Gamma}{2}M_s^2\right)^{\frac{\gamma+1}{2(\gamma-1)}}}.$$

Modeled is the computational regime of the flow in nozzle. The distribution function of the particles, flying into the nozzle, is:

$$f = \frac{n}{(\pi T)^{3/2}} \cdot \xi_x \exp\left(-\frac{c^2}{T}\right).$$

The reduction of macroparameters to a dimensionless form is realized on the basis of the condition at infinity. The concentration $n = 1$ corresponds to $N_* = 50$ particles in a cell. The field of flow is divided in 30 cells. To take into account the form of a nozzle it is proposed not to use the fact of collisions of the particles with the walls, because otherwise the problem will turn into two-dimensional one. It is possible to interpret the problem as one concerning the flow in the pipe of the constant cross section with porous walls. By means of blow-in and suction through the walls of that pipe one might create just such a flow which is realized in Lavalle nozzle of the prescribed geometry $S(x)$. Evidently, it is necessary to blow-in $N_i (1 - S(x))$ particles, where N_i is the number of particles in the ith cell.

In the present method all that was mentioned is realized in the following way. Introduced is the "weight" of particle, similar to that introduced by Bird. After that the calculation of macroparameters is realized as follows:

$$N_i^{j+1/2} = \sum_n \frac{S_{is,n}}{S_{i,n}} = \sum_n W_n,$$

$$\vec{u}_i^{j+1/2} = \frac{1}{N_i^{j+1/2}} \sum_n \vec{\xi}_n W_n,$$

$$Eu_c^{j+1/2} = \frac{1}{2} \sum_n \left(\vec{\xi}_{i,n} - \vec{u}_i^{j+1/2}\right) W_n,$$

$$E_t = \sum \frac{\gamma T_{is}}{4} \cdot W_n,$$

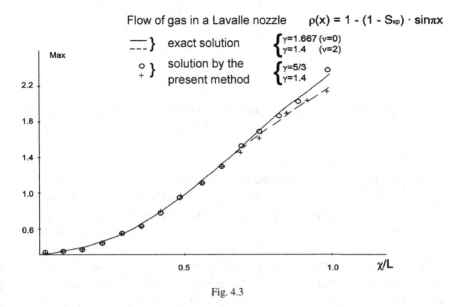

Fig. 4.3

$$E_i^{j+1/2} = E_c^{j+1/2} + E_t^{j+1/2},$$

$$p_i^{j+1} = \frac{4}{3 + \gamma} \frac{E_i^{j+1}}{N_i^{j+1/2}},$$

where $W_N = \frac{S_{is,n}}{S_{i,n}}$, S_{is} — the cross section in that cell, where the particle was located before the flight-over; S_i — the same after the flight-over.

The results of computations are presented in Fig. 4.3.

Chapter 5

Solution of the Navier–Stokes Equations (Petrov[133–139])

It seems that the numerical modeling of the flows of a continuous medium with the help of "fluids" (liquid particles) was initiated by Harlow.[144] The method to be proposed below differs principally both from the traditional finite-difference methods of solution of the Navier–Stokes equation, and from the classical method by Harlow, and by its internal nature is proves to be a synthetic of the method of statistical modeling (Monte Carlo method[131]) and of the method of discrete vortices (Belotserkovskii[132]). Here, one should mention the precursor of the present method — the classical work by Chorin,[146] from which the method indicated differs by its groundness.

Just as it was with the methods described in the preceding chapters, the problem should be stated in the form admitting the statistical modeling.

5.1 Formulation of the Problem, Initial and Boundary Conditions for the Navier–Stokes Equations in the Form by Helmholtz

Solution of the nonstationary problems for the Navier–Stokes equations proves sometimes to be simpler than the stationary ones. However, in the great bulk of practically important cases the researches are interested in the determination of the stationary or of the averaged over sufficiently large interval of time aerodynamical characteristics. Evidently, by the large interval of averaging vanishes the dependence of the result on the initial conditions, which could be taken in sufficiently arbitrary way. In the present formulation of a problem, it would be convenient to take the initial conditions in the form of the field of velocities appearing by a sudden setting of the body into motion from the state of rest.

In the general case, the problem of a flow of the incompressible fluid about the planar body is reduced to a solution of the nonstationary, two-dimensional Navier–Stokes equations, which in the rectilinear Cartesian coordinate system have the following form:

$$\frac{\partial V_x}{\partial t} + V_x \frac{\partial V_x}{\partial x} + V_y \frac{\partial V_x}{\partial y} = -\frac{1}{\rho_\infty} \frac{\partial p}{\partial x} + \nu \Delta V_x,$$

$$\frac{\partial V_y}{\partial t} + V_x \frac{\partial V_y}{\partial x} + V_y \frac{\partial V_y}{\partial y} = -\frac{1}{\rho_\infty} \frac{\partial p}{\partial y} + \nu \Delta V_y, \qquad (5.1.1)$$

$$\operatorname{div} \bar{V} = 0.$$

Here ν is a constant coefficient of kinematical viscosity, $\Delta = \frac{\partial^2}{\partial x^2} + \frac{\partial^2}{\partial y^2}$ — operator by Laplace, $\vec{V} = (V_x, V_y)$ — vector of velocity, p — static pressure, ρ_∞ — density of a medium, and t — time.

The part of initial condition for system (5.1.1) in some prescribed moment of time $t = \tau$ will be played by the prescribed field of velocities of a fluid $\bar{V}(\tau, \vec{r})$ corresponding to the instantaneous setting of the body to the motion with velocity $\vec{V}_\infty$ from the state of rest. It is assumed that in this case at the initial moment of time near the body is created a field of velocities for the break-away-free potential flow $V_0(\tau, \vec{r})$.[139]

At the contour of the body in flow L^*, at any moment of time (including the initial moment) should be fullfilled the condition of nonslip, which in the coordinate system connected with a body might be written down as

$$\vec{V}(t, \vec{r} \in L^*) = 0.$$

The field of velocities should also meet the boundary condition at infinity:

$$\vec{V}(t, \vec{r} \to \infty) = \vec{V}_\infty.$$

There is a tremendous amount of works on the numerical solution of the problems described by the Navier–Stokes equations for both incompressible fluid and compressible gas, in both the planar and spatial cases; at the present moment the listing of such works is already senseless.

After the introduction of the notion of vorticity of the flow,

$$\Omega_z = \Omega = \operatorname{rot} \vec{V}(t, \vec{r}),$$

the system of Eq. (5.1.1) and the problem as a whole might be reduced to the form

$$\frac{\partial \Omega}{\partial t} + V_x \frac{\partial \Omega}{\partial x} + V_y \frac{\partial \Omega}{\partial y} = \nu \Delta \Omega, \qquad (5.1.2)$$

$$\Omega = \operatorname{rot} \vec{V}; \quad \operatorname{div} \vec{V} = 0.$$

The systems of Eqs. (5.1.2), which presents in itself an extension of the Helmholtz's equations to the case of a viscous medium, is widely used for the study of flows of the viscous, incompressible fluid. Usually these equations are named as Navier–Stokes equations in the form by Helmholtz (sometimes "in the form by Gromeko-Lamb"). One of the advantages of this system, as compared with initial one, is the absence between the unknown functions of the pressure of a fluid. From the other hand, the flow's vorticity appears as in the certain sense natural variable for the problems of viscous flow, because just the essence of the viscous flow lies in the appearance of the vorticity within the boundary layer at the body and further, at the distance commensurable with the body's dimensions.

If by the numerical solution the setting of boundary conditions (nonslip condition) for the initial system (5.1.1) of Navier–Stokes equations does not present any special difficulties, then the setting of the same conditions for system (5.1.2) meets the certain complications. The principal difficulty by the numerical solution is created by the prescription of boundary conditions for the flow's vorticity. As a matter of fact, the prescription of the value of vorticity at the body's boundary is equivalent to the prescription of the surface friction, which is not known *a priori*, and for the determination of which, in particular, is solved the problem formulated.

At the other hand, the boundary condition for the vorticity is generally absent by the mathematically correct formulation of the problem, because in the presence of such a condition the problem becomes to be overdetermined. This leads to such a situation when the value of vorticity at the body's boundary is, usually, prescribed only approximately, and there are the special works devoted to the various manners of the formulation of such a prescription. Eventually, by the approximate manner is determined the surface tension at the boundary of a body. All this means that there exists a certain artificiality in the prescription of the boundary condition for vorticity, while the absence of that condition by the mathematically correct formulation of the problem leads to a suggestion of the physical ungroundness of the introduction of a condition mentioned. We are going to provide a number of the auxiliary proofs needed for a correct construction of the method proposed.

5.2   The General Properties of the Vertical Flow Arising by the Instantaneous Start of a Body from the State of Rest

Taking into account the importance of the question of the boundary conditions for vorticity, let us dwell more widely both on this matter and on the general laws of existence of a vertical flow, because the developed below method of solution of Eq. (5.1.2) does indirectly demand the observation of these laws.

Equations (5.1.2) are written down in the form

$$\frac{\partial \Omega}{\partial t} + \operatorname{div} \vec{Q} = 0,$$

$$\vec{Q} = \Omega \cdot \vec{V} - \nu \operatorname{grad} \Omega, \qquad (5.2.1)$$

$$\operatorname{div} \vec{V} = 0, \quad \Omega = \operatorname{rot} \vec{V}.$$

The introduced here vector $\vec{Q}$ might be treated as the density of the flow of vorticity, while the first of the equations (5.2.1) is just the continuity equation for vorticity, analogous to the usual continuity equation

$$\frac{\partial \rho}{\partial t} + \operatorname{div} \vec{G}_m = 0,$$

where $\vec{G}_m$ is the density of the flow of mass, $Q^* = \oint_{L*} (\vec{Q} \cdot \vec{n}) ds$.

As a preliminary step let us find the flow of vorticity through the unified contour $(L^* + L)$ with a cut R, where L^* is the contour of the body in the flow and L — the distant contour embracing the body (Fig. 5.1). It is evident that during the finite interval of time the vorticity is not able to move far of the body, to the infinity, and the initial (starting) vortices are located at some large, but finite distance $l = l_0$ away of the body. By moving of the external contour of integration L sufficiently far away, $l \gg l_0$ (almost infinity), and guessing that the vorticity at this contour is already absent, after the transformation with the help of a theorem by Green and taking into account the nonslip boundary condition, one obtains

$$Q^* = -\nu \oint_{L*} \frac{\partial \Omega}{\partial n} ds. \qquad (5.2.2)$$

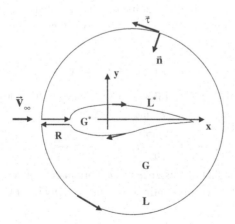

Fig. 5.1 The united contour of integration.

For the further transformation of expression for the flow of vorticity we are going to act in the following way. One needs to write down the expression for the pressure's drop between the two arbitrary points A and B at the contour of the body in the flow,

$$p_B - p_A = \int_A^B \left(\frac{\partial p}{\partial x} dx + \frac{\partial p}{\partial y} dy \right).$$

After the proper transformations with the use of Navier–Stokes equations (5.1.1) and the condition of nonslip one obtains

$$\frac{1}{\rho}(p_B - p_A) = \nu \int_A^B \frac{\partial \Omega}{\partial n} ds.$$

Since there should not be any breaks of pressure in the incompressible fluid, then by the complete pass around the contour we have $p_A = p_B$ and from here it follows

$$\nu \oint_{L^*} \frac{\partial \Omega}{\partial n} ds = 0. \tag{5.2.3}$$

Comparing this result with (5.2.2), one obtains

$$\oint_{L^*} (\vec{Q} \cdot \vec{n}) ds = 0. \tag{5.2.4}$$

Thus, we have shown that by the conditions prescribed the total flow of vorticity through the contour of the body in flow is constant and equal to zero. This condition might be treated as an integral boundary condition for vorticity, which should be observed always and by any chosen approach to the solution of system (5.1.2). For the observation of the integral boundary condition for vorticity without violation of the other conditions it would be sufficient to guess that there is a continuity of the density of vorticity's flow by the passing over the boundary L^* of the body in flow:

$$\vec{Q}(t, \vec{r} \in L^* + 0) = \vec{Q}(t, \vec{r} \in L^*) = \vec{Q}(t, \vec{r} \in L^* - 0).$$

Similarly, from relation (5.2.3) follows the continuity of the normal derivatives of vorticity by the passing over the boundary,

$$\frac{\partial \Omega}{\partial n}(t, \vec{r} \in L^* + 0) = \frac{\partial \Omega}{\partial n}(t, \vec{r} \in L^*) = \frac{\partial \Omega}{\partial n}(t, \vec{r} \in L^* - 0).$$

And, finally, from the fact of existence of the normal derivatives of vorticity at the points of a contour follows the continuity of the vorticity itself by the passing over the boundary,

$$\Omega(t, \vec{r} \in L^* + 0) = \Omega(t, \vec{r} \in L^*) = \Omega(t, \vec{r} \in L^* - 0). \tag{5.2.5}$$

The condition obtained might be treated in the following way: *formally, for the vorticity the body's boundary is absent, and the vorticity might freely cross this boundary, obeying only the integral conditions (5.2.3) and (5.2.4). Actually, as concerns the vorticity, the problem (5.1.2) might be considered as the Cauchy's problem.*

The fulfillment of the single boundary condition — that of nonslip — might be transposed on the solution of the problem of a search of the vorticity's distribution corresponding to that condition. The presence of the conditions of continuity (5.2.5) serves for the explanation of the absence in the totality of boundary conditions for the problem (5.1.2) of the other, physically grounded boundary condition for the vorticity.

Let us find one more integral condition for the vorticity, closely related to the preceding one. Aiming to that, one might integrate Eq. (5.1.2) over the area G, which forms the interior of the contour $(L^* + L)$:

$$\frac{\partial}{\partial t} \iint_G \Omega(x, y, t)dxdy = -\iint_G \operatorname{div}\vec{Q}dxdy.$$

After the application to the transformation of a right-hand side of the expression obtained of the theorem by Ostrogradskii and Gauss and upon shifting of the contour L into infinity and taking into account the boundary conditions both at the body and in infinity, one obtains

$$\frac{\partial}{\partial t} \iint_G \Omega(x, y, t)dxdy = 0,$$

or, as a consequence of that,

$$\iint_G \Omega(x, y, t)dxdy = \text{const.} \tag{5.2.6}$$

The same integral is computed over the area G^*, the interior of the body's contour L^*, because, as a consequence of continuity condition (5.2.5), the vorticity might, formally, cross the boundary of a body, thus creating a kind of flow in the interior of a contour. For this singly coherent area we have

$$\iint_{G^*} \Omega(x, y, t)dxdy = \iint_{G^*} \operatorname{rot}\vec{V}dxdy = \oint_{L^*} (\vec{V}d\vec{s}).$$

Due to the nonslip condition, which is also observed at the inner surface of a contour the last of the integrals above is equal to zero, identically. And then one has

$$\iint_{G^*} \Omega(x, y, t)dxdy = 0. \tag{5.2.7}$$

This expression might be interpreted as the equality to zero of the total intensity of vertical pipes inside the body's contour.

Let us determine the constant entering expression (5.2.6). To carry this out, let us transform the expression taking into account the boundary conditions both at the body and at infinity. One obtains

$$\iint_G \Omega(x, y, t)dxdy = \iint_G \mathrm{rot}\vec{V}dxdy = \oint_{L+L^*} (\vec{V}d\vec{s}) = \oint_L (\vec{V}d\vec{s}) = \Gamma_\infty,$$
(5.2.8)

where Γ_∞ is the circulation over the infinitely distant contour.

If one considers the problem of instantaneous start of the body from the state of rest, then it is clear that at the initial moment of time around the body is formed non-break-away and irrotational flow. Evidently, in this case the velocity's circulation $\Gamma_\infty(t = \tau)$ over the infinitely distant contour will be equal to zero. However, the integral in (5.2.6) is independent of time and, consequently, it will always be identically equal to zero,

$$\iint_G \Omega(x, y, t)dxdy = 0.$$
(5.2.9)

Thus, we have proved the validity of the following statement:

By the instantaneous start of the body in a flow of viscous, incompressible fluid the integral of arising vorticity over all the space is constant and equal to zero at any moment of time.

Or, otherwise: *the total flow of vorticity from the body's boundary and into the space is constant and equal to zero.*

Physically, it means that by the flow about body the vortices are always born in pairs, having the opposite signs, and their total intensity is always equal to zero. However, in the general case, when the initial conditions do not correspond to the instantaneous start of a body from the state of rest, the quantity $\Gamma_\infty(t = \tau) \neq 0$, and observed is the more general condition (5.2.8).

5.3   Initial Conditions for the Problem of the Instantaneous Start of a Body in a Viscous Fluid

As it was noted above, at the moment $t = \tau$, corresponding to the start of a body form the state of rest, formed near the body is a velocity field of a potential, irrotational and non-break-away flow. However, the effects of viscosity are revealed now within an infinitely thin layer of fluid at the body's surface, and is observed the nonslip condition, which included in itself also the condition of a nonflow-through. By the passing over this layer appears the breaking of a tangential component of velocity, which is equal to the velocity $V_0(\tau, s)$ of the irrotational, potential flow at the external boundary of a body, because at the inner boundary of the layer the velocity of flow equals to zero. Such a character of velocity field corresponds to the infinitely thin vertical shroud of the intensity $\gamma(\tau, s) = -V_0(\tau, s)$ pressed against the surface of a body. Methods of solution of the problem of determination of velocity $V_0(\tau, s)$ are developed quite well, and its determination for the arbitrary planar body does not meet any serious difficulties.

Taking into account the existence of breaking of the tangential component of velocity and of the nonslip condition, the velocity field in the immediate proximity of the wall, at the initial moment of time, might be presented as

$$V_n(\tau, s, n) = 0, \quad V_\tau(\tau, s, n) = V_0(\tau, s, n)H(n). \tag{5.3.1}$$

Here V_n and V_τ are the normal and the tangential components of velocity, correspondingly, n — the distance from the body's boundary in normal direction, $H(n)$ is the Heaviside function, defined by the condition

$$H(n) = \begin{cases} 0 & n < 0, \\ 1 & n > 0. \end{cases}$$

Using the presentation given for the velocity field, one might obtain the distribution of vorticity at the initial time moment:

$$\Omega(\tau, s, n) = \frac{\partial V_n}{\partial s} - \frac{\partial V_\tau}{\partial n} = -V_0(\tau, s)\frac{\partial}{\partial n}H(n).$$

The derivative of the function $H(n)$ is the δ-function by Dirac, and for that reason the final value of vorticity at the initial time moment τ might be presented in the form

$$\Omega(\tau, s, n) = -V_0(\tau, s)\delta(n).$$

The expression obtained indicated that the vorticity at the boundary of a body at the initial time moment belongs to the class of the generalized functions, and in accordance to the properties of δ-function is, strictly at the body's surface, equal to infinity.

The distribution of vorticity at the initial time moment might be, with the use of a δ-function of vectorial argument, presented as

$$\Omega_0(\tau, \vec{r}) = -\oint_{L^*} V_0(\tau, s) \cdot \delta[\vec{r} - \vec{r}(s)]ds. \tag{5.3.2}$$

Such a distribution of vorticity corresponds to the vertical layer of the intensity $\gamma(\tau, s) = -V_0(\tau, s)$, located at the body's surface.

In the computational methods the initial distribution of vorticity (5.3.2) will be conveniently presented by the system of discrete vortices. To achieve that one changes the right-hand side, approximately, for the integral sum:

$$\Omega_0(\tau, \vec{r}) \approx -\sum_{i=1}^{N} V_0(\tau, s_i)\delta[\vec{r} - \vec{r}(s_i)]\Delta s_i,$$

where N is the number of elements, into which the body's contour is divided, Δs_i — the element of contour's division.

The expression obtained corresponds to a system of discrete vortices, having the intensities

$$\Gamma_i = -V_0(\tau, s_i)\Delta s_i,$$

and located at the points s_i of the surface of a body in the flow.

As a final expression, the initial distribution of vorticity for computational methods in the form of a system of discrete vortices might be presented as

$$\Omega_0(\tau, \vec{r}) \approx -\sum_{i=1}^{N} \Gamma_i\delta[\vec{r} - \vec{r}(s_i)]. \tag{5.3.3}$$

5.4 The General Algorithm of the Numerical Solution of an Initial–Boundary Problem for the Navier–Stokes Equations in the form by Helmholtz

The obtained in a preceding paragraph initial and boundary conditions for vorticity, as well as its integral properties, permit to formulate the general principle of a solution of the initial–boundary problem for the Navier–Stokes equations in the form of Helmholtz. After the introduction in analysis the stream function of a flow the problem might be formulated as follows:

It is necessary to find the functions $\Omega(t, \vec{q})$, $\vec{V}(t, \vec{q})$, $\Psi(t, \vec{q})$, $\vec{q} = (x, y)$,

$$[\tau < t < T, \ T < \infty, \ |\vec{q}| < \infty, \ \Omega(t, \vec{q}) \in C_2,$$
$$\vec{V}(t, \vec{q}) \in C_3, \ \Psi(t, \vec{q}) \in C_2],$$

which satisfy the equations

$$\frac{\partial\Omega}{\partial t} + V_x\frac{\partial\Omega}{\partial x} + V_y\frac{\partial\Omega}{\partial y} = \nu\Delta\Omega,$$

$$\Delta\Psi = -\Omega, \ V_x = \frac{\partial\Psi}{\partial y}, \ V_y = -\frac{\partial\Psi}{\partial x}, \tag{5.4.1}$$

and also the initial and boundary conditions

$$\Omega(t = \tau, \vec{q}) = \Omega_0(\tau, \vec{q}), \quad \Psi(t, \vec{q} \in L^*) = 0, \quad \frac{\partial\Psi(t, \vec{q} \in L^*)}{\partial n} = 0.$$

Solution of the initial–boundary problem formulated is reduced, actually, to the unified solution of a Cauchy's problem for vorticity and a boundary problem for the stream function. Taking into account the results of a preceding paragraph, one might propose the following algorithm of solution of the problem stated.

Let us consider that at the initial moment of time $t = \tau$ both the velocity field in the whole surrounding space and the distribution of vorticity in the form of a vertical layer (5.3.2) are known. Then during the small interval of time Δt, considering the velocity field to be prescribed and invariable, let us solve the Cauchy problem for vorticity without taking into consideration the boundary conditions of nonslip and nonflow-through, imposed on the stream function. After that, let us find the distribution of vorticity $\Omega(t = \tau + \Delta t, \vec{q})$ in the space, at the moment of time $t = \tau + \Delta t$. Having the new distribution of vorticity, let us find the new velocity field in the presence of a body, which might be determined by the well-known methods and sufficiently simply, after taking into account the condition of nonflow-through, $\Psi(t, \vec{q} \in L^*) = 0$.

The condition of being equal to zero for the tangential component of velocity $V_\tau(\vec{q} \in L^*)$ at the body's contour, which is exactly fulfilled at the initial time moment, will be violated by the moment of time $t = \tau + \Delta t$, because the velocity field is assumed to be incorrected to correspond to that condition at the small interval of time. This means that at the moment of time $t = \tau + \Delta t$ at the surface of a body appears a certain "velocity of slip" — $V_s(\Delta t, s)$, the quantity characterizing the accuracy of observance of the nonslip velocity. It is assumed and, later on, will be proved that by such an algorithm of solution the time moment $t = \tau + \Delta t$ and all the subsequent moments do not, in principle, differ in any kind from the initial moment $t = \tau$. Then for the correction of the violated condition of nonslip, just as it was done at the first moment of time, one locates at the body's surface the vertical layer of the intensity $-V_s(\Delta t, s)$, which would correspond to the vorticity distribution,

$$\Omega_0(\tau + \Delta t, \vec{r}) = -\oint_{L^*} V_s(\tau + \Delta t, s) \cdot \delta[\vec{r} - \vec{r}(s)] ds.$$

It is clear that by locating at the body's surface of the vortical layer with intensity corresponding to the "velocity of slip" one would, once again, exactly observe the nonslip condition. And similar actions will be done at all the subsequent moments of time. The algorithm of such a type does, in some degree, reflect the process of the vorticity generation at the surface of a body and permits to achieve the exact fulfillment of the nonslip condition at the discrete moments of time which are considered as initial ones for each of those time intervals, at which the Cauchy problem is solved. In the first limit of $\Delta t \to 0$ the nonslip condition will be fulfilled exactly at each step in time and during the whole temporal interval.

Thus the algorithm described consists, actually, of two problems which are subsequently solved at each of the time steps. The first of these is the Cauchy problem for the equations by Helmholtz, and the second one is that of determination of the velocity field in the presence of a body, by the known distribution of vorticity in the space.

Solution of the Cauchy problem for the Navier–Stokes equations in the form by Helmholtz appears to be the most complicated and the most crucial part of the algorithm described because just in this part is revealed and taken into account the nature of viscous effects, for the sake of which the problem is solved. The problem itself is formulated in such a way. At the small step in time it is necessary to find the function $\Omega(t, \vec{q})$

$$[\tau < t < \tau + \Delta t, \quad \Delta t \to 0, \quad |\vec{q}| < \infty, \quad \Omega(t, \vec{q}) \in C_2],$$

which should satisfy the equation

$$\frac{\partial \Omega}{\partial t} + V_x \frac{\partial \Omega}{\partial x} + V_y \frac{\partial \Omega}{\partial y} = \nu \Delta \Omega, \tag{5.4.2}$$

and the initial condition

$$\Omega(\tau, \vec{r}) = - \oint_{L^*} V_0(\tau, s) \cdot \delta[\vec{r} - \vec{r}(s)] ds.$$

The components of the velocity vector $\vec{V} = (V_x, V_y)$ might be considered as known and invariable at the small temporal step. Actually, this assumption leads to a linearization of the problem formulated, and it is converted into the Cauchy problem for the linear differential equation of the parabolic type, and this conversion notably simplifies the process of solution.

To be sure, the problem's linearization at the small temporal step does not signify the linearization of the problem as a whole. At each interval of time consisting of the finite number of steps, solution of the problem as a whole will be nonlinear, because the velocity field is correct at each elementary step of time, and at the body's surface is generated a new vorticity.

Let us begin the solution of the Cauchy problem (5.4.2) formulated above. Using the linearity of that problem, one might put into consideration the fundamental solution of the Cauchy problem at the small period of time $\Delta t = t - \tau$ as a certain, temporarily unknown function of the four arguments, $f(\tau, \vec{p}, t, \vec{q})$, with the help of which the general solution of the problem is presented in the form

$$\Omega(t, \vec{q}) = \iint_G \Omega(\tau, \vec{p}) f(\tau, \vec{p}, t, \vec{q}) d\xi d\eta,$$
$$\vec{q} = (x, y), \quad \vec{p} = (\xi, \eta). \tag{5.4.3}$$

The fundamental solution has quite definite physical sense. If for the part of an initial distribution of the vorticity is taken the point vortex of the unit intensity, located initially at the point $\vec{p}_0$, then the quantity $\Omega(\tau, \vec{p})$ might be presented in the form

$$\Omega(\tau, \vec{p}) = \delta(\vec{p} - \vec{p}_0).$$

Substituting this form into the general solution (5.4.3), one obtains

$$\Omega(t, \vec{q}) = \iint_G \delta(\vec{p} - \vec{p}_0) f(\tau, \vec{p}, t, \vec{q}) d\xi d\eta = f(\tau, \vec{p}_0, t, \vec{q}).$$

Thus, when speaking on the physical sense, the fundamental solution $f(\tau, \vec{p}, t, \vec{q})$ defines the law of motion and of the diffusion of a point vortex of unit intensity in the flow of the viscous incompressible fluid.

Introduction of the function $f(\tau, \vec{p}, t, \vec{q})$ permits to pass from the problem of a search of the vorticity distribution $\Omega(t, \vec{q})$ to a problem of looking for the universal fundamental solution, and this last problem is usually simpler. When the fundamental solution is obtained, the initial condition turns into the general one; more determinated properties of the fundamental solution lead, as a rule, to the essential narrowing of the class of functions within which the solution of a problem stays as possible.

There is yet one more consideration in favor of passing to the search of a fundamental solution. It is known that a fundamental solution of the linear equations of the parabolic type possesses the properties of the density of probability for some accidental Markovian process.

As soon as one takes into consideration the properties of vorticity described earlier, one is able to show formally that the introduced in such a way function $f(\tau, \vec{p}, t, \vec{q})$ is really possessing all the properties of the density of probability for an accidental Markovian process. Substituting the general solution (5.4.3) into the Helmholtz equation (5.4.2), one obtains

$$\iint_G \Omega(\tau, \vec{p}) \left[\frac{\partial f}{\partial t} + V_x \frac{\partial f}{\partial x} + V_y \frac{\partial f}{\partial y} - \nu \Delta f \right] d\xi d\eta = 0.$$

Due to the arbitrariness of the initial distribution of vorticity $\Omega(\tau, \vec{p})$ for the observation of the condition obtained it would be sufficient to set

$$\frac{\partial f}{\partial t} + V_x \frac{\partial f}{\partial x} + V_y \frac{\partial f}{\partial y} = \nu \Delta f. \tag{5.4.4}$$

Thus for the introduced function $f(\tau, \vec{p}, t, \vec{q})$ one obtains just the same linear differential equation of a parabolic type. From the initial condition for vorticity follows the initial condition for the function $f(\tau, \vec{p}, t, \vec{q})$:

$$f(\tau, \vec{p}, t = \tau, \vec{q}) = \delta(\vec{q} - \vec{p}). \tag{5.4.5}$$

It should be noted that the initial condition for $f(\tau, \vec{p}, t, \vec{q})$ appears to be universal for all the problems of the flow of viscous incompressible fluid about the body.

One of the main properties of parabolic equations of the type of (5.4.4) with positive factors at the second derivatives is the property of the leveling off in time of the values of the function sought in different points of the space. In the particular case of our initial condition one of the consequences of the property mentioned will appear the positiveness of the function $f(\tau, \vec{p}, t, \vec{q})$ at any arbitrary moment of time t:

$$f(\tau, \vec{p}, t, \vec{q}) > 0, \quad \tau < t < \infty.$$

Let us obtain some integral properties of the function introduced. After integration of (5.4.2) over all the space one obtains

$$\iint_{G+G^*} \Omega(t, \vec{q}) dx dy = \iint_G \Omega(\tau, \vec{p}) \left[\iint_{G+G^*} f(\tau, \vec{p}, t, \vec{q}) dx dy \right] d\xi d\eta.$$

For the observation of conditions (5.2.7), (5.2.9) and on account of the initial condition (5.4.5) for the function $f(\tau, \vec{p}, t, \vec{q})$ it would be sufficient to set

$$\iint_{G+G^*} f(\tau, \vec{p}, t, \vec{q}) dx dy = 1.$$

This means that the function $f(\tau, \vec{p}, t, \vec{q})$ should observe the condition of the rate setting.

Let us obtain one more integral property of the function $f(\tau, \vec{p}, t, \vec{q})$ connecting its values at various moments on time. Let us consider three moments on time in succession $\tau \to \tau + \Delta\tau \to t$ and three corresponding to them distributions of vorticity, which would successively passing one into another, $\Omega(\tau, \vec{q}) \to \Omega(\tau + \Delta\tau, \vec{q}) \to \Omega(t, \vec{q})$. The expressions for those are obtained from the general solution (5.4.2):

$$\Omega(\tau + \Delta\tau, \vec{q}) = \iint_G \Omega(\tau, \vec{p}) f(\tau, \vec{p}, \tau + \Delta\tau, \vec{q}) d\xi d\eta, \tag{5.4.6}$$

$$\Omega(t, \vec{q}) = \iint_G \Omega(\tau + \Delta\tau, \vec{z}) f(\tau + \Delta\tau, \vec{z}, t, \vec{q}) dz_1 dz_2, \tag{5.4.7}$$

$$\vec{z} = (z_1, z_2).$$

From the other hand, eluding the intermediate moment of time, $\tau + \Delta\tau$, from (5.4.2) one obtains

$$\Omega(t, \vec{q}) = \iint_G \Omega(\tau, \vec{p}) f(\tau, \vec{p}, t, \vec{q}) d\xi d\eta. \tag{5.4.8}$$

The expression (5.4.6) for $\Omega(\tau + \Delta\tau, \vec{q})$ is substituted into (5.4.7) and by comparison with (5.4.8) one obtains

$$f(\tau, \vec{p}, t, \vec{q}) = \iint_G f(\tau, \vec{p}, \tau + \Delta\tau, \vec{z}) \cdot f(\tau + \Delta\tau, \vec{z}, t, \vec{q}) dz_1 dz_2. \qquad (5.4.9)$$

The expression obtained presents in itself the integral equation connecting the values of a function $f(\tau, \vec{p}, t, \vec{q})$ at various time moments. Its form coincides with that of integral equation by Smoluchowski, which is well known in the statistical physics.[140] *The Smoluchowski equation is satisfied by the densities of probability of transition for the systems passing during their development through the sequence of states forming the chain by Markov.* The processes described by this equation are named as *the processes of Markovian type*, or the processes without afteraction. Taking into account all the obtained properties of the function $f(\tau, \vec{p}, t, \vec{q})$ — positiveness, the condition of rate setting, the form of initial condition, — one might firmly state that the introduced above fundamental solution of the Cauchy problem for Navier–Stokes equations possesses all the properties of the density of probability for some accidental Markovian process.

This peculiarity of the fundamental equation introduced, $f(\tau, \vec{p}, t, \vec{q})$, opens by its determination the way to the extensive use of the methods, nontraditional for aerodynamics, namely — those of the theory of probability and, in particular, methods of the theory of the accidental Markovian processes and statistical modeling.

For the correct setting of the problem of looking for the fundamental solution $f(\tau, \vec{p}, t, \vec{q})$ it would be necessary, moreover, to formulate the boundary conditions agreed upon the integral boundary conditions for the vorticity of a flow (5.2.3). From the condition of continuity of the flow of vorticity at the boundary and of continuity of the vorticity itself one will be able without particular difficulties to obtain

$$f(\tau, \vec{p}, t, \vec{q} \in L^* - 0) = f(\tau, \vec{p}, t, \vec{q} \in L^*) = f(\tau, \vec{p}, t, \vec{q} \in L^* + 0),$$

$$\oint_{L^*} \frac{\partial f}{\partial n} ds = 0.$$

These boundary conditions are usually interpreted as the continuity of the density of probability by the transition of the boundary and continuity of the flow of the density of probability. By the observation of the conditions of diffusional isotropy (in the present case — constancy of the coefficient of viscosity both in space and in time), these conditions prove to be the traditional conditions for the density of probability for the accidental Markovian processes.

Let us obtain one additional property of the function $f(\tau, \vec{p}, t, \vec{q})$, which would simplify the procedure of search for this function. Equation (5.4.4) defines dependence of the fundamental solution on arguments $t, \vec{q}$, but application of this equation is, quite evidently, insufficient for a definition of $f(\tau, \vec{p}, t, \vec{q})$ as a function of four arguments. It was shown (see Kolmogorov[141]), that there is a possibility to obtain, from Eq. (5.4.9) by Smoluchowski, a system of two differential equations:

$$\frac{\partial f}{\partial t} + V_x(t, \vec{q})\frac{\partial f}{\partial x} + V_y(t, \vec{q})\frac{\partial f}{\partial y} = \nu\Delta f,$$

$$\frac{\partial f}{\partial \tau} + V_x(\tau, \vec{p})\frac{\partial f}{\partial \xi} + V_y(\tau, \vec{p})\frac{\partial f}{\partial \eta} = -\nu\Delta f. \quad (5.4.10)$$

The equations obtained are in the general case named as *Kolmogorov equations*. Equations of this system are conjugated between them, and, as a consequence of this, their solution will depend on the difference of the corresponding arguments, only,

$$f(\tau, \vec{p}, t, \vec{q}) = f(t - \tau, \vec{q} - \vec{p}).$$

Thus for a definition of $f(\tau, \vec{p}, t, \vec{q})$ it would be sufficient to use only one (the first) equation of the system (5.4.10).

The first of Kolmogorov's equations is usually called as equation by Fokker–Plank.[142,143]. Ultimately, the problem of a search for the fundamental equation might be formulated in such a way:

One should find the function $f(\tau, \vec{p}, t, \vec{q})$,

$$[\tau < t < \tau + \Delta t, \ \Delta t \to 0, \ |\vec{q}| < \infty, \ f(\tau, \vec{p}, t, \vec{\vec{q}}) > 0 \in C_2,$$
$$f = f(t - \tau, \vec{q} - \vec{p})],$$

which satisfies the Fokker–Plank equation

$$\frac{\partial f}{\partial t} + V_x\frac{\partial f}{\partial x} + V_y\frac{\partial f}{\partial y} = \nu\Delta f, \quad (5.4.11)$$

an initial condition

$$f(\tau, \vec{p}, t = \tau, \vec{q}) = \delta(\vec{q} - \vec{p}),$$

conditions of the continuity at the boundary, and of the rate setting

$$f(\tau, \vec{p}, t, \vec{q} \in L^* - 0) = f(\tau, \vec{p}, t, \vec{q} \in L^*) = f(\tau, \vec{p}, t, \vec{q} \in L^* + 0),$$

$$\iint_{G+G^*} f(\tau, \vec{p}, t, \vec{q})dxdy = 1.$$

We were going to prove, earlier, that all the moments of time are on equal terms for the general problem, and all the temporal steps are realized similarly to the first one. Actually, this statement is already justified, for the property of that type is just the peculiar quality of the systems, which in their development pass through the sequence of states, forming the chain by Markov. The state of Markovian system at the time moment $t = \tau + \Delta t$ is completely defined by its state at the preceding moment of time τ, and is independent of the preceding states. This property permits to solve the problem independently and uniformly at each step in time.

Ultimately, taking into account introduction of the fundamental solution, it is proposed to use the following algorithm of a solution of the problem (5.4.1) for the instantaneous start of a body within the incompressible viscous fluid.

The whole interval of time needed for us is divided into N steps Δt, and each of those is considered as small. At each of the steps in time the solution might be subdivided in three stages. At the first stage the field of velocities is determined in space and at the body's boundary, staying on account of a possible presence in the space of some vorticity $\Omega(\tau, \vec{p})$ and on account of the condition of nonflow-through at the body. For this aim solved is the linear boundary problem for the stream function

$$\Delta \Psi = -\Omega(\tau, \vec{p}),$$
$$\Psi(\tau, \vec{p} \in L^*) = 0. \qquad (5.4.12)$$

As a result of the solution of this problem is found the tangential component of velocity at the body's surface, $V_s(\tau, s)$ — "velocity of slip." It is to be noted here, that at the strictly first step in time the vorticity in space is wittingly absent, and the "velocity of slip" is equal to the velocity of potential, non-break-away and irrotational flow.

Further on, at the surface of a body is located an initial distribution of vorticity (vertical layer), which might be presented in the form

$$\Omega(\tau, \vec{p}) = -\oint_{L^*} V_s(\tau, s) \cdot \delta[\vec{p} - \vec{p}(s)]ds. \qquad (5.4.13)$$

It is to be noted, that now is formally observed the nonslip condition, and here the first stage is over.

At the second stage solved is the linear Cauchy problem for the Fokker–Plank equation, and it is assumed, that the field of velocities is known at the small intervals of time as a result of the problem's solution at the first stage,

$$\frac{\partial f}{\partial t} + V_x \frac{\partial f}{\partial x} + V_y \frac{\partial f}{\partial y} = \nu \Delta f, \qquad (5.4.14)$$
$$f(\tau, \vec{p}, t = \tau, \vec{q}) = \delta(\vec{q} - \vec{p}).$$

As a result of this solution defined is the function $f(\tau, \vec{p}, \tau + \Delta t, \vec{q})$, a knowledge of which permits to realize a transition to the third stage of solution and to find the spatial vorticity distribution $\Omega(\tau + \Delta t, \vec{q})$:

$$\Omega(\tau + \Delta t, \vec{q}) = \iint_G \Omega(\tau, \vec{p}) f(\tau, \vec{p}, \tau + \Delta t, \vec{q}) d\xi d\eta. \qquad (5.4.15)$$

This means that the solution of problem (5.4.1) is built up at the small interval of time Δt.

At the next step in time one should once again return to the problem for a stream function (5.4.12), to find the new velocity field in space and the "velocity of slip" at the body's surface. One proceeds to locate at the body the vertical layer (5.4.13), solves the Cauchy problem (5.4.14), finds the vorticity distribution (5.4.15) and goes on until the achievement of time of solution needed for the obtainment of some aerodynamical characteristics.

The proposed above algorithm of solution contains two main problems. The first of them is the linear boundary problem for the stream function (5.4.12) which does not, as a rule, meet any principal difficulties.

The second main problem, the search for a fundamental solution of the Navier–Stokes equations, i.e. the function $f(\tau, \vec{p}, t, \vec{q})$, is reduced to a linear Cauchy problem for the Fokker–Plank equation (5.4.14). The next paragraph of this chapter will be devoted to the solution of this problem with the help of the methods of theory of probability, taking into account all the peculiarities imposed by the general hydrodynamical character of the problem considered.

5.5   Solution of the Cauchy Problem for the Fokker–Plank Equation at Small Interval of Time

As it was shown above, the search of the fundamental solution for Helmholtz equation at the small interval of time is reduced to a solution of the linear Cauchy problem for the Fokker–Plank equation

$$\frac{\partial f}{\partial t} + V_x \frac{\partial f}{\partial x} + V_y \frac{\partial f}{\partial y} = \nu \Delta f, \tag{5.5.1}$$

$$f(\tau, \vec{p}, t = \tau, \vec{q}) = \delta(\vec{q} - \vec{p}).$$

The general methods of solution of this problem are sufficiently well developed in the classical theory of probabilities.[26,27] It is known also that the effectiveness of the application of one or another method of solution of that problem is strongly dependent on the concrete form and on the properties of the convectional and dispersional functions. As applied to the hydrodynamics, the Fokker–Plank equations were not, apparently, used, and the proper methods of their solution were not developed till the publication of works by Petrov,[135,136] where for the solution of a problem stated is proposed to use the method of moments and the method of statistical modeling (Monte Carlo method), based on the transition from the Fokker–Plank equation to the statistically equivalent system of the stochastic differential equations.

The choice of the methods mentioned is not a casual one. The method of moments permits to build up with a prescribed degree of accuracy the approximate analytical solution of the problem and to analyze its properties within one or another area of flow. The moments of distribution function have, themselves, a certain physical sense and might be used for the active analysis of the behavior of vorticity. Method of the statistical modeling is more convenient for the realization on the electronic computer, and, at the same time, is closely connected with well developed for the ideal fluid method of the discrete vortices and with Lagrangian approach to the mechanics of continuum.

Before the transition to the solution of the Cauchy problem (5.5.1) with the use of the methods chosen above, let us consider some particular problems having the exact analytical solution. The analysis of these will give a possibility to get acquainted with the concrete form of the function $f(\tau, \vec{p}, t, \vec{q})$ and with its properties.

Let us consider the following problem. Assume that the fluid is moving along the concentric circles around the coordinate's origin. The motions of such a type appear by the diffusion of the arbitrary axisymmetrical distribution of vorticity in the boundless space. In this case, the Cauchy problem for the Fokker–Plank equation will have a form

$$\frac{\partial f}{\partial t} = \nu \left(\frac{1}{r} \frac{\partial f}{\partial r} + \frac{\partial^2 f}{\partial r^2} \right),$$

$$f(\tau, \rho, t = \tau, r) = \delta(r - \rho).$$

The equation obtained looks as a classical equation of the heat conduction, possessing the well known unique solution, which satisfies the initial condition, the condition of positiveness and of the rate setting,

$$f(\tau, \rho, t, r) = \frac{1}{4\pi\nu(t - \tau)} \exp\left[-\frac{(r - \rho)^2}{4\nu(t - \tau)} \right]. \tag{5.5.2}$$

Let the initial condition for vorticity correspond to the point vortex of the intensity Γ, located at the origin,

$$\Omega(\tau, \rho) = \Gamma\delta(\rho).$$

Then, with the help of (5.4.2) one obtains the general solution of the corresponding Cauchy problem for the vorticity of flow:

$$\Omega(t, r) = \int_0^\infty \frac{\Gamma\delta(\rho)}{4\pi\nu(t - \tau)} \exp\left[-\frac{(r - \rho)^2}{4\nu(t - \tau)} \right] 2\pi\rho d\rho.$$

Conducting the integration with account for the properties of δ-function, one obtains the well-known exact solution,[28] corresponding to a diffusion of the point vortex:

$$\Omega(t, r) = \frac{\Gamma}{4\pi v(t - \tau)} \exp\left[-\frac{r^2}{4v(t - \tau)}\right].$$

The exact solution (5.5.2), obtained above, shows that in the particular case considered function $f(\tau, \vec{p}, t, \vec{q})$ presents in itself the Gaussian (normal) function of distribution of the density of probability with the mathematical expectation $\bar{r} = \rho$ and with dispersion $\sigma = 2v(t - \tau)$, which prove to be the first initial and second central moments of the distribution function, correspondingly.

Let us begin to build up the solution of the Cauchy problem (5.5.1) in the general case. Taking into account the extremal abundance in the nature of the normal Gaussian distribution, one will look for the general solution of the problem in the form of two-dimensional Gaussian distribution function, which is determined, in general case, by the six parameters — two mathematical expectations $\bar{x}$, $\bar{y}$, and four moments of the second order — D_{xx}, D_{yy}, D_{xy}, D_{yx}:

$$f(\tau, \vec{p}, t, \vec{q}) = \frac{1}{2\pi\sqrt{D_{xx}D_{yy}(1 - r^2)}}$$

$$\times \exp\left\{-\frac{1}{2(1 - r^2)}\left[\frac{(x - \bar{x})^2}{D_{xx}} - 2r\frac{(x - \bar{x})(y - \bar{y})}{D_{xy}} + \frac{(y - \bar{y})^2}{D_{yy}}\right]\right\},$$

$$r^2 = \frac{D_{xy}^2}{D_{xx}D_{yy}}. \tag{5.5.3}$$

The moments of the distribution function are not yet known and are subject to the determination. In the two-dimensional case, one has

$$\frac{\partial f}{\partial t} + \frac{\partial}{\partial x}(V_x f) + \frac{\partial}{\partial y}(V_y f) = v\Delta f, \quad \operatorname{div}\vec{V} = 0, \tag{5.5.4}$$

$$f(\tau, \vec{p}, t = \tau, \vec{q}) = \delta(\vec{q} - \vec{p}).$$

As usually, one introduces the mathematical expectations or the mean (over the space) values:

$$\bar{A}(t, \vec{p}) = \iint_G A(t, \vec{q}) f(\tau, \vec{p}, t, \vec{q}) dx dy, \tag{5.5.5}$$

where $A(t, \vec{q})$ is an arbitrary function of the coordinates and of time.

Let us assume, moreover, that this integral always converges, or, at any rate, for the real physical problems, in which it is usually assumed that exist the limits

$$f(\tau, \vec{p}, t, \vec{q} \to \infty) = 0,$$

$$x^m y^n \frac{\partial^{m+n}}{\partial x^m \partial y^n} f(\tau, \vec{p}, t, \vec{q} \to \infty) = 0. \tag{5.5.6}$$

To obtain the equations for initial moments of the first order, or, otherwise, — for the mathematical expectations of the function $f(\tau, \vec{p}, t, \vec{q})$, one should multiply Eq. (5.5.4), successively, by $(x - \bar{x})$ and $(y - \bar{y})$, and to integrate the result over the whole space. Taking into account the notations introduced and the properties of (5.5.6) one obtains the differential equations for the mathematical expectations of the function $f(\tau, \vec{p}, t, \vec{q})$:

$$\frac{\partial \bar{x}(t, \vec{p})}{\partial t} = \iint_G V_x(t, \vec{q}) f(\tau, \vec{p}, t, \vec{q}) dx dy = \bar{V}_x(t, \vec{p}),$$

$$\frac{\partial \bar{y}(t, \vec{p})}{\partial t} = \iint_G V_y(t, \vec{q}) f(\tau, \vec{p}, t, \vec{q}) dx dy = \bar{V}_y(t, \vec{p}).$$

After integration of the equations obtained over the time, one obtains the expressions for the mathematical expectations in quadratures,

$$\bar{x}(t) = \xi + \int_\tau^t \bar{V}_x(t, \vec{p}) dt,$$

$$\bar{y}(t) = \eta + \int_\tau^t \bar{V}_y(t, \vec{p}) dt. \tag{5.5.7}$$

As an immediate sequence of the expressions obtained it is seen that the mathematical expectations of the function $f(\tau, \vec{p}, t, \vec{q})$, coinciding with its maximum, are transported together with a fluid along the flowlines, with some mean velocity of a flow.

The equations for all the central moments of the second and of the higher order will be obtained in the general form. To obtain them, Eq. (5.5.4) is multiplied by $(x - \bar{x})^n (y - \bar{y})^m$ and integrated over the whole space, taking into account the properties (5.5.6). After some transformations one obtains

$$\frac{\partial D_{nx,my}(t, \vec{p})}{\partial t}$$

$$= \iint_G n(x - \bar{x})^{n-1}(y - \bar{y})^m V_x(t, \vec{q}) f(\tau, \vec{p}, t, \vec{q}) dx dy$$

$$+ \iint_G m(y - \bar{y})^{m-1}(x - \bar{x})^n V_y(t, \vec{q}) f(\tau, \vec{p}, t, \vec{q}) dx dy$$

$$+ \nu \iint_G n(n - 1)(x - \bar{x})^{n-2}(y - \bar{y})^m f(\tau, \vec{p}, t, \vec{q}) dx dy$$

$$+ \nu \iint_G m(m - 1)(x - \bar{x})^n (y - \bar{y})^{m-2} f(\tau, \vec{p}, t, \vec{q}) dx dy. \tag{5.5.8}$$

Here $D_{nx,my} = \iint_G (x - \bar{x})^n (y - \bar{y})^m f(\tau, p, t, q) dx dy$.

The initial conditions for (5.5.7) and (5.5.8) are presented as

$$\bar{x}(t = \tau) = \xi, \quad \bar{y}(t = \tau) = \eta, \quad D_{nx,my}(t = \tau) = 0.$$

Let us write down the complete system of equations for the moments, up to the moments of the second order, necessary for the construction of the function $f(\tau, \vec{p}, t, \vec{q})$ in the Gaussian approximation:

$$\frac{\partial \bar{x}(t, \vec{p})}{\partial t} = \iint_G V_x(t, \vec{q}) f(\tau, \vec{p}, t, \vec{q}) dx dy,$$

$$\frac{\partial \bar{y}(t, \vec{p})}{\partial t} = \iint_G V_y(t, \vec{q}) f(\tau, \vec{p}, t, \vec{q}) dx dy,$$

(5.5.9)

$$\frac{\partial D_{xx}}{\partial t} = 2v + 2 \iint_G V_x(t, \vec{q}) f(\tau, \vec{p}, t, q)(x - \bar{x}) dx dy,$$

$$\frac{\partial D_{yy}}{\partial t} = 2v + 2 \iint_G V_y(t, \vec{q}) f(\tau, \vec{p}, t, q)(y - \bar{y}) dx dy,$$

$$\frac{\partial D_{xy}}{\partial t} = \iint_G V_x(t, \vec{q}) f(\tau, \vec{p}, t, q)(y - \bar{y}) dx dy$$

$$+ \iint_G V_y(t, \vec{q}) f(\tau, \vec{p}, t, q)(x - \bar{x}) dx dy,$$

$$D_{xy} = D_{yx}.$$

The equations obtained for the determination of moments are exact ones, but their direct solution is yet presently impossible, because in the right-hand parts of these equations appears the unknown function $f(\tau, \vec{p}, t, \vec{q})$. Usually one acts in the following way. The convectional function (velocity field) is expanded in the Taylor series in vicinity of the mathematical expectation, i.e. the point $\vec{p}_0 = (\bar{x}, \bar{y})$. By doing that assumed is the sufficient smoothness of the velocity field and the differentiability of that field,

$$V_x(t, \vec{q}) = V_x(t, \vec{p}_0) + \frac{\partial V_x(t, \vec{p}_0)}{\partial x}(x - \bar{x}) + \frac{\partial V_x(t, \vec{p}_0)}{\partial y}(y - \bar{y}) + \cdots,$$

$$V_y(t, \vec{q}) = V_y(t, \vec{p}_0) + \frac{\partial V_y(t, \vec{p}_0)}{\partial x}(x - \bar{x}) + \frac{\partial V_y(t, \vec{p}_0)}{\partial y}(y - \bar{y}) + \cdots.$$

(5.5.10)

Carrying out the substitution of expansions obtained into (5.5.9) and limiting oneself with the initial moments of the first and second order, one obtains the set of differential equations of the first order:

$$\frac{\partial \bar{x}}{\partial t} = V_x(t, \vec{p}_0) + \frac{1}{2}\frac{\partial^2 V_x(t, \vec{p}_0)}{\partial x^2}D_{xx} + \frac{1}{2}\frac{\partial^2 V_x(t, \vec{p}_0)}{\partial y^2}D_{yy} + \frac{\partial^2 V_x(t, \vec{p}_0)}{\partial x \partial y}D_{xy},$$

$$\frac{\partial \bar{y}}{\partial t} = V_y(t, \vec{p}_0) + \frac{1}{2}\frac{\partial^2 V_y(t, \vec{p}_0)}{\partial x^2}D_{xx} + \frac{1}{2}\frac{\partial^2 V_y(t, \vec{p}_0)}{\partial y^2}D_{yy}$$

$$+ \frac{\partial^2 V_y(t, \vec{p}_0)}{\partial x \partial y}D_{xy}, \tag{5.5.11}$$

$$\frac{\partial D_{xx}}{\partial t} = 2v + 2\frac{\partial V_x(t, \vec{p}_0)}{\partial x}D_{xx} + 2\frac{\partial V_x(t, \vec{p}_0)}{\partial y}D_{xy},$$

$$\frac{\partial D_{yy}}{\partial t} = 2v + 2\frac{\partial V_y(t, \vec{p}_0)}{\partial x}D_{xy} + 2\frac{\partial V_y(t, \vec{p}_0)}{\partial y}D_{yy},$$

$$\frac{\partial D_{xy}}{\partial t} = \frac{\partial V_x(t, \vec{p}_0)}{\partial y}D_{yy} + \frac{\partial V_y(t, \vec{p}_0)}{\partial x}D_{xx}.$$

Before the beginning of the search for solution of this set at the small time interval $\Delta t = t - \tau$, one estimates the accuracy of the Gaussian approximation. As it is known, the degree of distinction from the normal distribution is determined by the third central moments — the excesses. Let us write down the equation for determination of the excess along the axis OX:

$$\frac{\partial D_{xxx}}{\partial t} = 3 \iint_G (x - \bar{x})^2 V_x(t, \vec{q}) f(\tau, \vec{p}, t, \vec{q})dxdy.$$

Let us estimate the order of the quantity D_{xxx}. Substituting expansion (5.5.10) for the velocity field and carrying out the integration over the small time interval $\Delta t = t - \tau$, one obtains

$$D_{xxx} \approx 6vV_x(\tau, \vec{p}_0)\Delta t^2 + O\left(\frac{\Delta t^3}{Re}\right),$$

where Re is the Reynolds number.

Similarly, for the excess D_{yyy} one obtains

$$D_{yyy} \approx 6vV_y(\tau, \vec{p}_0)\Delta t^2 + O\left(\frac{\Delta t^3}{Re}\right).$$

Thus the Gaussian approximation of the function $f(\tau, \vec{p}, t, \vec{q})$ is valid within the whole area of flow, with the accuracy up to the terms of the order $O(\frac{\Delta t^2}{Re})$. The exception is formed only by those areas of flow, where the components of the velocity vector are small or equal zero. For example, such a situation exists at the wall, in the area of a boundary layer, due to the nonslip condition. There the Gaussian approximation is valid with accuracy, equal, at least, up to the terms of order $O(\frac{\Delta t^3}{Re})$. Apparently, directly at the wall the Gaussian approximation is valid just strictly, and the indication for such a validity is made by the external form of the equations studied with account on the nonslip condition. And as soon as one is aware that the area of the wall and of the boundary layer appears as the most important from the point of view concerning the revelation of the viscous effects, one might consider the choice of the Gaussian approximation as the solution of the problem (5.5.1) to be sufficiently justified.

Let us write down the solution of a system (5.5.11) in the form of series in time and preserving only the terms of the order of $O(\Delta t^2)$ and $O(\frac{\Delta t}{Re})$. Let us name such a solution as Gaussian approximation of the first order:

$$\bar{x}(t) = \xi + V_x(\tau, \vec{p}_0)\Delta t + \frac{dV_x(\tau, \vec{p}_0)}{dt}\frac{\Delta t^2}{2} + O(\Delta t^3) + O\left(\frac{\Delta t^2}{Re}\right),$$

$$\bar{y}(t) = \eta + V_y(\tau, \vec{p}_0)\Delta t + \frac{dV_y(\tau, \vec{p}_0)}{dt}\frac{\Delta t^2}{2} + O(\Delta t^3) + O\left(\frac{\Delta t^2}{Re}\right),$$

$$D_{xx}(t) = 2v\Delta t + O\left(\frac{\Delta t^2}{Re}\right), \qquad\qquad (5.5.12)$$

$$D_{yy}(t) = 2v\Delta t + O\left(\frac{\Delta t^2}{Re}\right),$$

$$D_{xy} = O\left(\frac{\Delta t^2}{Re}\right).$$

In this approximation the function $f(\tau, \vec{p}, t, \vec{q})$ acquires the simple and understandable appearance:

$$f(\tau, \vec{p}, t, \vec{q}) = \frac{1}{4\pi v(t-\tau)} \exp\left[-\frac{(x-\bar{x})^2 + (y-\bar{y})^2}{4v(t-\tau)}\right]$$

$$+ O(\Delta t^3) + O\left(\frac{\Delta t^2}{Re}\right). \qquad\qquad (5.5.13)$$

In the frame of just that approximation one might write down the general solution of the problem (5.4.1) at the time interval $\Delta t = t - \tau$:

$$\Omega(t, \vec{q}) = \iint_G \frac{\Omega_0(\tau, \vec{p})}{4\pi v(t-\tau)} \exp\left[-\frac{(x-\bar{x})^2 + (y-\bar{y})^2}{4v(t-\tau)}\right]d\xi d\eta$$

$$+ O(\Delta t^3) + O\left(\frac{\Delta t^2}{Re}\right). \qquad\qquad (5.5.14)$$

Here $\Omega_0(\tau, \vec{p})$ is the initial distribution of vorticity, which has its own form for each step in time.

Let us analyze the solution obtained. If the part of an initial distribution of the vorticity at the moment of time $t = \tau$ is played by the point vortex located at the point $\vec{p}_0 = (\xi, \eta)$, then up to moment of time $t = \tau + \Delta t$ the maximum of the vorticity distribution is shifted with a velocity of flow and together with a fluid till the point of a mathematical expectation. Simultaneously, due to the influence of viscosity, occurs the diffusion of vorticity, and by the moment of time t the part of a characteristical size of vertical zone will be played by the value of dispersion of the distribution function, $D \approx 2\nu(t - \tau)$. One might say that in this approximation is exactly observed the principle by Harlow[144] concerning the splitting according to the physical processes. According to this principle and in application to the concrete case considered, the physical processes of the vorticity's transfer and of its diffusion occur independently of each other. In the next approximation, concerning the time, the principle of splitting on the physical processes is violated.

5.6 The Numerical Solution of the Fokker–Plank Equation by the Method of Direct Statistical Modeling

The obtained above solution (5.5.14) is, principally, closing the proposed here algorithm of solution of the initial–boundary problem for Navier–Stokes equations, which was introduced in Sec. 5.4. However, by the numerical realization of that algorithm, for the computation of the integral (5.5.14) it is necessary to divide all the space into the finite elements, and such a process is always sufficiently laborious. It is possible to propose yet another method of construction of the function $f(\tau, \vec{p}, t, \vec{q})$ and of the general solution of Cauchy problem basing on the method of moments without division of the space in cells. This method is closely connected with methods of direct statistical modeling (Monte Carlo methods) and with the Lagrangian approach to the mechanics of continuous medium.

It is known that in the case of positiveness and of scalar nature of a coefficients by the second derivatives in the Fokker–Plank equation (in the case under investigation such as always positive coefficient of viscosity) occurs the possibility of a mutually simple transition from the Fokker–Plank equation to the set of the ordinary, stochastical differential equations with the argument coinciding with the ordinate of accidental process. Such a way of action is extremely widely used by the description of diffusional Markovian processes obeying the equation of Fokker–Plank in a quite variable multitude of the areas of science. In all the cases, the part of density of the distribution of probability should be played by the function $f(\tau, \vec{p}, t, \vec{q})$.

For the case under consideration the set of stochastical differential equations, which are in statistical sense equivalent to the Fokker–Plank equations, has a form

$$\frac{d\vec{q}_j(t)}{dt} = \vec{V}(t, \vec{q}_j) + \sqrt{2\nu}\vec{\xi}(\vec{q}_j, t),$$

$$\vec{q}_j(t = \tau) = \vec{p}_i(\tau), \quad i, j = 1, 2, 3 \ldots, \tag{5.6.1}$$

$$\vec{q}_j(t) = (x_j, y_j).$$

Here $\vec{\xi}(\vec{q}_j, t) = (\xi_x, \xi_y)$ is the accidental vector function possessing the properties of a "white noise," $\vec{V}(t, \vec{q}_j)$ — vector of the mathematical expectation of the velocity (mean velocity) of a flow.

Here and further on we shall assume the *ergodic property* of the accidental process (5.6.1). This property of the accidental Markovian processes permits to find the mathematical expectations of the accidental values as the obtained by averaging in time. For example, introduced earlier mathematical expectation of the arbitrary function (5.5.5) with the use of assumption of the ergodic property might be expressed in the following way:

$$\bar{A}(t, \vec{p}) = \iint_G A(t, \vec{q}) f(\tau, \vec{p}, t, \vec{q}) dx dy = \frac{1}{T} \int_\tau^t A(\tau, \vec{p}) d\tau,$$

$$T = t - \tau.$$

Here the integration is conducted over the whole period T of the prehistory of the process development. For this reason in the further exposition the mathematical expectations will be interpreted as mean values in time.

The assumption of the ergodic property is frequently appearing in aerodynamics. For example, in the theory of turbulence is usually proposed the smoothing of the velocity field with the help of some "smooth-away function" possessing the proper properties, but later this function is forgotten and the velocity fields are smoothed just over time.

Let us transform the set of stochastical equations (5.6.1). For the realization of that this set is integrated in respect of time, in the interval $\Delta t = t - \tau$,

$$\vec{q}_j(t) = \vec{p}_i(\tau) + \int_\tau^t \vec{V}(t, \vec{q}_j) dt + \sqrt{2\nu} \int_\tau^t \vec{\xi}(\vec{q}_i, t) dt. \tag{5.6.2}$$

The integral of the accidental function $\vec{\xi}(\vec{q}, t)$ appears also to be a certain accidental function. It is denoted as $\vec{\xi}_0(\vec{q}_j, t)$:

$$\sqrt{2\nu} \int_\tau^t \vec{\xi}(\vec{q}_j, t) dt = \vec{\xi}_0(\vec{q}_j, t).$$

Now it will be demanded that the ordinate of the accidental process $\vec{q}_j(t)$ would be distributed with the density of probability $f(\tau, \vec{p}, t, \vec{q})$ determined by the Fokker–Plank equation. To achieve that it would be sufficient to demand that the moments $\vec{q}_j(t)$, which are understood now as the corresponding mean values of time, would coincide with the found earlier moments of the function $f(\tau, \vec{p}, t, \vec{q})$ as, for example, in the Gaussian approximation (5.5.12):

$$\overline{(x_j - \bar{x}_i)^2} = D_{xx}, \quad \overline{(y_j - \bar{y}_i)^2} = D_{yy},$$
$$\overline{(x_j - \bar{x}_i)(y_j - \bar{y}_i)} = D_{xy} = D_{yx}. \tag{5.6.3}$$

Since from now on the moments of the process (5.6.2) are determined by an accidental function $\vec{\xi}_0(\vec{q}_j, t)$, let us demand that this function would be distributed in normal way. To achieve that this function is presented in the form

$$\vec{\xi}_0(\vec{q}_j, t) = \hat{S} \cdot \vec{\zeta}_0(t).$$

Here $\vec{\zeta}_0(t) = (\zeta_{0x}, \zeta_{0y})$ is an accidental vectorial function possessing the properties of "white noise", with noncorrelated components, distributed in normal way, having a zero mathematical expectation and a unit dispersion.

$\hat{S} = \begin{pmatrix} s_{xx} & s_{xy} \\ s_{yx} & s_{yy} \end{pmatrix}$ is a matrix which might be called as "diffusional" one and the terms of which should be determined. Properties of the function $\vec{\zeta}_0(t) = (\zeta_{0x}, \zeta_{0y})$ might be expressed in the following way:

$$\overline{\zeta_{0x}(t)} = \overline{\zeta_{0y}(t)} = 0, \quad \overline{\zeta_{0x}^2(t)} = 1, \quad \overline{\zeta_{0y}^2(t)} = 1,$$
$$\overline{\zeta_{0x}\zeta_{0y}(t)} = 0. \tag{5.6.4}$$

The expressions for the moments (5.5.12) contain the temporarily unknown initial moments of the first order, $\bar{x}_j(t)$, $\bar{y}_j(t)$. Going to find these moments, presently as the mean values in time, let us remark that in the Gaussian approximation (5.5.13) the vorticity's diffusion is isotropic. As a sequence of that, the accidental function $\vec{\xi}_0(\vec{q}_j, t)$ and the terms of a diffusional matrix $\hat{S}$ are independent of the point of space, but depend only on time. Then, realizing the averaging in time of Eq. (5.6.2) for $\vec{q}_j(t)$ and taking into account zero value of the mathematical expectation $\vec{\xi}_0(\vec{q}_j, t)$, one obtains in the component's form

$$\bar{x}_i(t) = \xi_i + \int_\tau^t V_x(\vec{q}_i, t)dt,$$
$$\bar{y}_i(t) = \eta_i + \int_\tau^t V_y(\vec{q}_i, t)dt. \tag{5.6.5}$$

The sign of temporal averaging for the components of the velocity field is deliberately omitted because the formulation of the problem assumes, from the very beginning, the laminar character of flow and the absence of the accidental component of the velocity field. It is easy to note that the mathematical expectations for the ordinates of accidental process (5.6.5), which were found from the stochastical equations (5.6.1), absolutely coincide within the frames of approximation admitted, with the same expectations obtained earlier from the equations of moments (5.5.7).

The physical sense of the expressions (5.6.5), presented above, is extremely clear and consists in the fact that the vertical particles of a fluid are transferred with the maximal probability (in the average measure), together with the fluid, along the flowlines, and in the approximation considered the transfer of vorticity is independent on its diffusion.

Now let us determine the terms of the diffusional matrix $\hat{S}$. For the realization of that let us present in more detailed form the expressions for moments (5.6.3), using in this process the introduced diffusional matrix $\hat{S}$:

$$\overline{(x_j - \bar{x}_i)^2} = \overline{(s_{xx}\zeta_{0x} + s_{xy}\zeta_{0y})^2}, \quad \overline{(y_j - \bar{y}_i)^2} = \overline{(s_{yx}\zeta_{0x} + s_{yy}\zeta_{0y})^2},$$

$$\overline{(x_j - \bar{x}_i)(y_j - \bar{y}_i)} = \overline{(s_{xx}\zeta_{0x} + s_{xy}\zeta_{0y})(s_{yx}\zeta_{0x} + s_{yy}\zeta_{0y})}.$$

After fulfillment of the temporal averaging of the function $\vec{\zeta}_0(t) = (\zeta_{0x}, \zeta_{0y})$ and taking into account the properties (5.6.4), one gets the equations serving to determination of the terms of a diffusional matrix:

$$s_{xx}^2 + s_{xy}^2 = D_{xx},$$

$$s_{yy}^2 + s_{yx}^2 = D_{yy},$$

$$s_{xx}s_{xy} + s_{yy}s_{yx} = D_{xy} = D_{yx}.$$

Solving this set of algebraical equations, one obtains the unknown terms of the diffusional matrix:

$$s_{xx} = \frac{D_{xx} + \sqrt{D}}{\sqrt{D_{xx} + D_{yy} + 2\sqrt{D}}},$$

$$s_{yy} = \frac{D_{yy} + \sqrt{D}}{\sqrt{D_{xx} + D_{yy} + 2\sqrt{D}}},$$

$$s_{xy} = \frac{D_{xy}}{\sqrt{D_{xx} + D_{yy} + 2\sqrt{D}}},$$

$$s_{xy} = s_{yx}, \quad D = D_{xx}D_{yy} - D_{xy}^2.$$

Using the obtained earlier values of moments (5.5.12) in the Gaussian approximation of the first order, one finds the terms of a matrix $\hat{S}$ in their explicit form:

$$\hat{S} = \begin{pmatrix} \sqrt{2v(t-\tau)} & 0 \\ 0 & \sqrt{2v(t-\tau)} \end{pmatrix} + O\left(\frac{\Delta t^{\frac{3}{2}}}{Re}\right).$$

Now let us describe in the approximation chosen the accidental process, which
stays in the mutually unique correspondence with a solution of the Cauchy problem
for the Fokker–Plank equation, and the function of distribution of the density of
probability $f(\tau, \vec{p}, t, \vec{q})$ in the Gaussian approximation (5.5.13):

$$x_j(t) = \xi_i + \int_\tau^t V_x(\vec{q}_i, t)dt + \sqrt{2v(t-\tau)}\zeta_{0x}(t) + O\left(\frac{\Delta t^{\frac{3}{2}}}{Re}\right),$$

$$y_j(t) = \eta_i + \int_\tau^t V_y(\vec{q}_i, t)dt + \sqrt{2v(t-\tau)}\zeta_{0y}(t) + O\left(\frac{\Delta t^{\frac{3}{2}}}{Re}\right).$$

$$(5.6.6)$$

The vectorial form of these expressions is:

$$\vec{q}_j(t) = \vec{p}_i(\tau) + \int_\tau^t \vec{V}(\vec{q}_i, t)dt + \sqrt{2v(t-\tau)}\vec{\zeta}_0(t) + O\left(\frac{\Delta t^{\frac{3}{2}}}{Re}\right).$$

Just like it was the case with a solution (5.5.14), in the last expression is clearly
visible the principle of splitting on the physical processes for the transfer of vorticity
and its diffusion.

The expression, similar to that in its sense, was obtained directly on the basis
of the principle of splitting on the physical processes in the work by Chorin,[146]
and was used for the building-up of the computational scheme for the flow about
planar bodies with a small viscosity.

In the inviscid fluid the accidental terms in expressions (5.6.6) are absent, and
these expressions turn into the original expressions of the method of discrete vortices in the ideal fluid, in which the vortices are moving strictly along the flowlines.
Thus, the procedure of obtainment of the function $f(\tau, \vec{p}, t, \vec{q})$ with the help of an
accidental process (5.5.11) might be considered as an extension of the method of
discrete vortices on the case of the viscous fluid and as the variety of Monte Carlo
method.

Let us show how the accidental process (5.6.6) might be used for the building-
up of a function $f(\tau, \vec{p}, t, \vec{q})$. Let the area of flow as a whole is divided into totality
of finite elements (cells) having the area Δs_k (see Fig. 5.2).

Then, by definition, the quantity $f(\tau, \vec{p}_i, t, \vec{q}_j)\Delta s_j$ will appear as a probability
of entrapment of the ordinate of a process beginning at the time $t = \tau$, at the point
$\vec{p}_i = (\xi_i, \eta_i)$, by the cell "$j$" with a coordinate $\vec{q}_j$. From the other hand, by way of

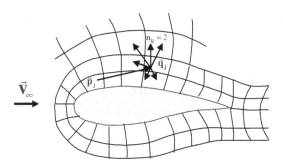

Fig. 5.2   The scheme of "gambling" of the accidental process (5.6.6).

"gambling" of the accidental process N times and by the counting of number n_{ij} of entrapment of the ordinate of process by the cell "j", one obtains

$$f(\tau, \vec{p}_i, t, \vec{q}_j)\Delta s_j \approx \frac{n_{ij}}{N}.$$

The exact value of the function $f(\tau, \vec{p}_i, t, \vec{q}_j)$ is obtained by means of "gambling" of the process (5.6.6) realized extremely large number of times and by means of tending to zero of the area of elements of division,

$$f(\tau, \vec{p}_i, t, \vec{q}_j) = \lim_{\substack{N \to \infty \\ \Delta s_j \to 0}} \frac{n_{ij}}{N} \frac{1}{\Delta s_j}.$$

After the change of integral for the corresponding sum, the general solution of the problem for the flow's vorticity in the Gaussian approximation will look as follows:

$$\Omega(t, \vec{q}_j) = \lim_{\substack{N \to \infty \\ \Delta s_{i,j} \to 0}} \sum_{i=1}^{N_0} \Omega(\tau, \vec{p}_i) \frac{n_{ij}}{N} \frac{\Delta s_i}{\Delta s_j}, \tag{5.6.7}$$

where N_0 is the number of cells by the division of velocity field.

If one passes from the notion of vorticity to that of the circulation of velocity over the cell's contour, then one might impart to the general solution the following appearance:

$$\Gamma(t, \vec{q}_j) = \lim_{N \to \infty} \sum_{i=1}^{N_0} \Gamma(\tau, \vec{p}_i) \frac{n_{ij}}{N}. \tag{5.6.8}$$

The physical sense of a discrete (or numerical) form of the general solution is sufficiently clear. Let us assume that the viscosity of a flow tends to zero (an ideal fluid) and the accidental term vanishes. Then by the gambling of the process (5.6.6) its ordinate will always be entrapped by one and the same cell, and will be observed the condition $n_{ij} = N$. In such a case, the vorticity is transferred without diffusion, and the discrete vortices are moving strictly along the flowlines. If the fluid is viscous one, then the presence of the accidental term leads to such a situation, when the ordinate of the accidental process, being entrapped with a maximal probability by one and the same cell, just as in ideal fluid, but, after all, comes also into the neighboring cells. By that process for any cell will be observed the condition $n_{ij} < N$, and the absolute values of vorticity in a cell and of a circulation over its contour will diminish. Consequently, realized is the diffusion of vorticity.

For the confirmation of validity of the theoretical premises admitted above and of the efficiency of a numerical algorithm proposed, was conducted a numerical solution of the Cauchy problem for the diffusion of a point vortex within the viscous fluid staying in rest at the infinity. The accidental process (5.6.6) was realized in accordance to the following algorithm:

$$x_j(t) = \xi_i(\tau) + V_x(\tau, \vec{p}_i)\Delta t + \frac{dV_x(\tau, \vec{p}_i)}{dt}\frac{\Delta t^2}{2} + \sqrt{2v(t-\tau)}\zeta_{0x}(t),$$

$$y_j(t) = \eta_i(\tau) + V_y(\tau, \vec{p}_i)\Delta t + \frac{dV_y(\tau, \vec{p}_i)}{dt}\frac{\Delta t^2}{2} + \sqrt{2v(t-\tau)}\zeta_{0y}(t), \quad (5.6.9)$$

$$\vec{p}_i(\tau) = (\xi_i, \eta_i).$$

The process (5.6.9) was "gambled" $N = 2000$ times with a temporal step $\Delta t = 0.01/v$, and was counted the number of entrapments of the ordinates of a

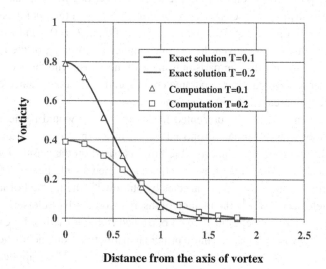

Distance from the axis of vortex

Fig. 5.3

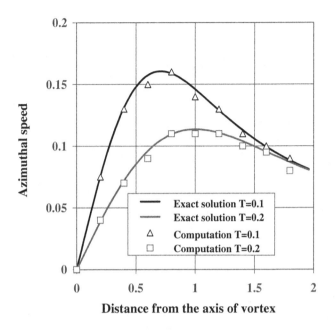

Fig. 5.4

process by the ring $r_i < r < r_i + \Delta r$. After that, the value of vorticity was obtained with the help of expression (5.6.7). The field of velocities was calculated by the ordinary way, as from the ensemble of discrete vortices in the boundless flow of an ideal fluid. In Figs. 5.3 and 5.4, by two values of time: $T = 0.1$ and $T = 0.2$ (in the fractions of $1/\nu$), are given the obtained values of vorticity, of the tangential velocity of a flow, and the comparison is made of these data with those of the exact solutions known. By the chosen parameters of solution the coincidence of results should be considered as a satisfactory one.

Summarizing the results presented in this chapter, it would be necessary to stress the following positions. Proposed by Petrov the methodics of a statistical modeling of the viscous incompressible flows is based on the classical works on the numerical modeling of the flows of continuous media with the help of "fluids" (liquid particles): the method of particles by Harlow,[141] the method of large particles by Belotserkovskii,[12] the method of discrete vortices by Belotserkovskii,[132] and with the use of a stochastic element borrowed from the method by Chorin.[146] The approach by Petrov has one more principal distinction from the methods mentioned above, and it consists of the careful groundness and of the applicability to the solution of aerodynamical problems.

Chapter 6

Studies of the Weakly Perturbed Flows
of Rarefied Gas

The numerous phenomena of the dynamics of rarefied gas, which have the practical and methodological value, possess the weakly nonequilibrium character and are described by the linearized kinetic equation. The totality of these phenomena includes the majority of the problems on flows within the thin layers near the walls, the problems of a motion of the bodies in gas with small velocities, of the determination of the coefficients of transfer and of the boundary conditions for the flows of continuous media, and some other problems. Reduced to the problems of that kind are also the essentially nonequilibrium flows, when the main parameters of such a flow are known from some other considerations and subject to determination are the small deviations from these solutions. Just for the problems of that kind were developed the methods proposed in Chap. 2 and applied here to the solution of the series of practically important problems.

6.1 Determination of the Velocity of Slip

As it is known, the explanation by Gilbert–Enskog–Chapman gives the solutions asymptotically converging to that of the Boltzmann equation by the Knudsen numbers tending to zero. However, near the boundaries, by as small as one wishes Knudsen number exists the area, within which the series mentioned above does not present the solution of Boltzmann equation. The thickness of this area, which is named as Knudsen layer, is of the order of a length of free path λ. Let us consider the flow of gas, which stays above the rigid plane wall, by the presence of a gradient of the mass speed, which at the far distance from the wall is assumed to be constant. Let us choose the coordinate system as it is shown in Fig. 6.1, where NS denotes the area of Navier–Stokes flow, K — the area of Knudsen.

The aim of the present work consists in the determination of such fictitious macroscopical boundary conditions for Navier–Stokes equations at the solid wall, by the observation of which the solution of Navier–Stokes equations outside of the Knudsen layer would coincide with the solution of Boltzmann equation with the prescribed real kinetic conditions at the wall.

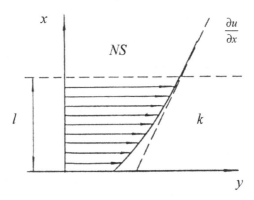

Fig. 6.1

For simplicity it is assumed that at the surface of a body the molecules flying upwards are subject to the law of a diffuse reflection,

$$f_{00} = n_0 \left(\frac{m}{2\pi k T_0} \right)^{3/2} \exp \left(-\frac{m}{2k T_0} \xi^2 \right). \qquad (6.1.1)$$

At the infinity (practically — at the distance of mean free path) is prescribed the distribution function by Navier–Stokes,

$$f = f_0 \left[1 + \frac{P_{xy}}{2\rho} \frac{m}{2kT} C_x C_y \right], \quad f_0 = f_{00}[1 + 2C_y u(x)h_0],$$

$$\bar{C} = \bar{\xi} - \bar{u}, \quad P_{xy} = -\mu \left(\frac{\partial u}{\partial x} \right)_\infty, \quad h_0 = \frac{m}{2kT_0}. \qquad (6.1.2)$$

Within the Knudsen layer the distribution function appears in the form

$$f = f_{00}[1 + 2h_0 \xi u(x) + \phi(x, \xi_x, \xi_x)]. \qquad (6.1.3)$$

For the solution of this problem will be used the method developed in Sec. 2.1 of the Chap. 2.

As it is seen from (6.1.2), the upper of the boundary conditions contains the value of a mass speed, which in the first approximation might be prescribed arbitrarily. Here was used a convenient algorithm of the determination of a mass speed at the upper boundary (see Khlopkov[98]). The field of flow is divided into a number of layers. Assuming that in vicinity of the upper boundary the profile of velocity acquires the prescribed gradient $(\partial u/\partial x)_\infty$ of Navier–Stokes type, the mass speed

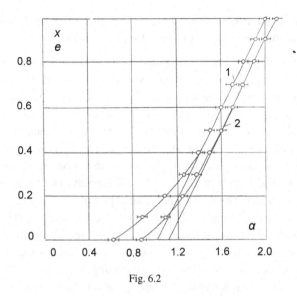

Fig. 6.2

at the upper boundary will be determined by the formula

$$u_i^{(n)} = u_{i-1}^{(n-1)} + \left(\frac{\partial u}{\partial x}\right)_\infty \Delta x, \qquad (6.1.4)$$

where i is the number of layer at the upper boundary, Δx — the thickness of a layer, n — the number of approximation.

Shown in Fig. 6.2 are the profiles of dimensionless velocity. Denoted by Fig. 6.1 is the profile of velocity for pseudo-Maxwellian molecules taken from the work by Gorelov and Kogan[102] (the coefficient of slip is equal to 1.015), by Fig. 6.2 is denoted the profile of velocity for the solid spheres, obtained in the work by Khlopkov[98] (the coefficient of slip equals to 1.103). As it is known, the Maxwellian molecules and solid spheres present in themselves two extreme cases of the potential of molecular interaction. For this reason, it would natural to assume that the results for the molecules with all the other potentials of interaction will be found within the intermediate area between the values indicated above.

In connection with technical complexities of the experimental modeling of the flows with Knudsen layers, the numerical computation was for a sufficiently long time the only source of information on the structure of such a layer. It is to be noted that the experiment on the determination of the velocity of slip, conducted abroad several years later (see Reynolds, Smolderen, Wendt[103]) confirmed the validity of results obtained and described here.

6.2　Solution of the Problem of the Feeble Evaporation (Condensation) from the Plane Surface (see Korovkin, Khlopkov[104])

With the help of Monte Carlo method, developed by Vlasov and Khlopkov[99] for the linear problems of the molecular gas dynamics was solved, among others, the problem of computation of the temperature gap within the Knudsen layer by the presence at the wall either condensation or evaporation. The similar problems were solved earlier either by way of finding the distribution function, or by way of solution of the model equation. The aim of the present study consists in development of the linear kinetic theory of condensation and evaporation by the small Knudsen numbers. Proceeding by this way is solved the linearized Boltzmann equation, and this solution is conducted by the original method permitting to realize the economy of computer memory. At the far distance from the boundary of the division of phases are valid the Navier–Stokes equations. Thus, the kinetical description of the boundary zone determines the conditions of the physical mating of the Navier–Stokes area with the boundary of the conversion of phases.

Let us remind the essence of the method applied. Let the size of a cell be equal to Δy, and then the probability of collision within that cell is equal to $\sigma_0 n \Delta t$, where $\Delta t = \Delta y / |\xi_y|$ denotes the interval of stay of the trial molecule within this cell. The probability of the realization of the flying of the molecule considered into the cell with a velocity $\vec{\xi}$ is proportional to $|\xi_y| f(\vec{\xi})$. Then the probability of collision of the trial molecule within that cell is proportional to

$$|\xi_y| f \sigma_0 n \Delta t = |\xi_0| f \sigma_0 n \frac{\Delta y}{|\xi_y|} \sim f(\vec{\xi}), \qquad (6.2.1)$$

while the probability of the fact, that the particle colliding with the trial molecule, possesses the velocity $\vec{\xi}$, is also proportional to $f(\vec{\xi})$. Consequently, the computation might be conducted in such a way: within the cell, where the collision occurs, one needs to memorize the velocity, with which the particle flies into this cell, while the computation of the collision one needs to conduct with that velocity, which was memorized after the preceding collision.

6.2.1　*Setting of the problem*

In the coordinate system (x, y) is considered the steady process of condensation (evaporation) at the surface $y = 0$. The motion of gas towards the surface (or outwards of it) is realized with macroscopic speed $\vec{u} = (0, \vec{u}, 0)$ (the motion of gas along the surface is absent). Since considered is the process with a small degree of nonequilibrium, this process of condensation or evaporation will be "slow", that is, one has $u/C_W = \tilde{u} \ll 1$, where $C_W = (m/2kT)^{-1/2}$, and from here it follows that within the layer is permissible the linearization of the distribution function and of the hydrodynamical parameters.

$$f = f_0(1 + \varphi(y, \xi)),$$

$$f_0 = n_0 \left(\frac{m}{2\pi k T_0} \right)^{-3/2} \exp \left(-\frac{m\xi^2}{2kT_0} \right), \tag{6.2.2}$$

$$T = T_0(1 + \tau(y)), \quad n = n_0(1 + \nu(y)), \tag{6.2.3}$$

where f_0 is the equilibrium distribution function, $T_0 = T_W$ is the temperature of a surface, $n_0 = n_W$ is corresponding to it equilibrium density, $\phi(y, \vec{\xi})$, $\tau(y)$, $\nu(y)$ — small additions, of which the squares might be neglected ($\phi(y, \vec{\xi}) \ll 1, \tau(y) \ll 1, \nu(y) \ll 1$).

Let us assume that the molecules of gas interact with the surface according to the following model. Let at the surface $y = 0$ fall the molecules with intensity J_-. Let the part a_W of the molecules falling is "absorbed" by the wall, while the part $(1 - a_W)$ is diffusely reflected from the surface with the temperature of surface T_W. Moreover, the surface evaporates the gaseous particles with intensity J_+, which, according to the numerous experiments, depends only on the properties of the material of a wall and on its temperature, and this is also realized diffusely, with the temperature T_0. The coefficient a_W is determined experimentally. For the beginning will be considered the particular case, when $a_W = 1$, that is, when all the molecules, falling at the surface, are absorbed. Proposed here is the simple way of the recounting of thus obtained results for the general case. Such an approach permits to shorten the computational process through the decrease of the number of the problem's parameters. At the upper boundary of the Knudsen layer the distribution function acquires the form of Navier–Stokesian one (see Fig. 6.3).

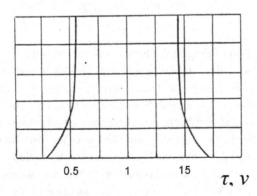

Fig. 6.3

6.3 The Slow Motion of a Sphere in Rarefied Gas (Brownian Motion)

The results of experimental measurement of the drag of a sphere in the flow of rarefied gas with a small velocity were published by Millican as long ago as in 1923; however, the attempts of a theoretical study of the sphere's drag in the transitional area extending from the regime of continuous medium till the free-molecular on, began to be executed only from the mid-1960s. For both of the limiting regimes are existing the analytical expressions for a force of drag.

In the first case (regime of continuum), the force of a drag is provided by the Stokes' theory. It is assumed that Mach number M tends to zero faster than the Knudsen number Kn; this assumption provides the conditions for a validity of the assumption by Stokes concerning the smallness of the Reynolds number Re:

$$Re = \frac{M}{Kn} \to 0. \tag{6.3.1}$$

As it is known, in this case the theory by Stokes gives for a diffuse reflection of molecules from the sphere's surface (nonslip condition) the value of the drag force, equal to $6\pi\mu Ru$, and for a specular reflection (condition of the zero value of the tangential momentum) the value, equal to $4\pi\mu Ru$, where R is the sphere's radius, μ — viscosity, u — speed of the oncoming flow.

In the second case, when $Kn \sim \infty$, the value of a drag force is provided by the formula of the free-molecular flow about a sphere. Coming to the limit with $u \to \infty$ one obtains

$$F_{0\rho} = \rho_\infty \frac{\pi R^2 u^2}{2S} \left[\frac{16}{3\sqrt{\pi}} + \frac{2}{3}\sqrt{\pi} \right], \tag{6.3.2}$$

for a diffuse reflection,

$$F_{03} = \rho_\infty \frac{\pi R^2 u^2}{2S} \frac{16}{3\sqrt{\pi}}, \tag{6.3.3}$$

for a specular reflection.

For a computation of drag in the transitional area it is necessary to solve the Boltzmann equation, and this is mating with the considerable difficulties of mathematical character, which are connected, from one side, with a complicated appearance of the integral of collisions, and, from the other side, with a multi-dimensional character of the problem. For that reason all the attempts of solution of the kinetical equation were limited either by the approximation of the distribution function, or by the analysis of Krook's equation. The special remark should be made concerning the work by Cercignani and his co-workers, where the variational method for the model equation by Krook is used for the computation of the drag of a sphere in the regimes of flow from one for continuous medium till the free-molecular one. The reflection of molecules from the sphere's surface was assumed by Cercignani to be diffuse. The result of computation showed the good correlation with the experimental data by Millican.

In the present exposition the sphere's drag is determined by way of solution of the Boltzmann equation with the help of Monte Carlo method (see Khlopkov[105]).

6.3.1 *Setting of the problem*

For the achievement of the aim indicated above let us consider the sphere of radius R located in the origin of a rectilinear coordinate system. The field of flow was bounded by the surface of a cube with its center located in the coordinate's origin and with its side equal to 10 calibers. In view of the symmetry of the problem considered was only the upper half of the flow, while at the plane $y = 0$ was prescribed a specular law of the reflection of molecules. At the infinity (at the sides of a cube) was prescribed the equilibrium distribution function with the parameters n_∞, u_∞, and T_∞, and the law of the reflection of molecules from the sphere's surface was assumed to be either specular, or diffuse with the temperature equal to that at infinity:

$$f_W = n_W \left(\frac{m}{2\pi k T_\infty} \right)^{3/2} \exp \left(-\frac{m}{2k T_\infty} \xi^2 \right). \tag{6.3.4}$$

The number by Knudsen is introduced by the usual way:

$$Kn = \frac{\lambda_\infty}{R}. \tag{6.3.5}$$

The Markovian chain represents in itself just roaming of the trial molecule about the equilibrium distribution function:

$1°$. The velocity of a molecule, which flies in from the boundary, is distributed proportionally to

$$e^{-v^2} (v_n). \tag{6.3.6}$$

$2°$. The duration of the free flight is gambled with a probability

$$e^{-k\tau} K_0 d\tau, \tag{6.3.7}$$

where

$$K_0 = a \int g e^{-v_1^2} dv_1. \tag{6.3.8}$$

$3°$. The velocity after collision is proportional to

$$e^{-v_1^2} g. \tag{6.3.9}$$

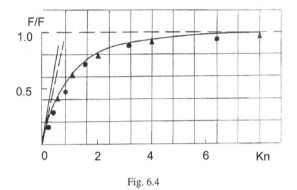

Fig. 6.4

The functional to be found is the drag force:

$$F_x = m \int v_z(v_n) f_0 \varphi \, dv \, ds, \qquad (6.3.10)$$

where the integration is carried out over the whole surface of a sphere. This quantity obtained is estimated as a mean arithmetic value of the accidental value $h = m v_x(v_n)$ at the point of the sphere's surface.

The results of computation of the drag force in relation to its value for the free-molecular flow are shown in dependence on the number Kn and presented in Fig. 6.4.

6.4 The Coefficient of Diffusion and the Mean Shifting of a Brownian Particle in the Rarefied Gas (see Khlopkov[106])

The phenomenon of a chaotic motion of the small particles, suspended within a fluid, served as a source for some works by Einstein, in which he obtains the formulae for the coefficient of diffusion D and the mean shifting λ_x:

$$D = \frac{kT}{K}, \qquad (6.4.1)$$

$$\lambda_x = \sqrt{2Dt}, \qquad (6.4.2)$$

where k is the Boltzmann's constant, T is the temperature of a fluid, K is a coefficient of proportionality, and t — time.

As the main assumption here is considered the small velocity of the motion of a particle, and from this assumption follows that the medium's resistance is directly proportional to the particle's velocity v:

$$F = Kv. \tag{6.4.3}$$

To obtain the concrete values of the coefficient of diffusion and of the mean shifting it would be necessary to know the value of the coefficient K. Following Einstein, the coefficient of proportionality is usually found from the formula by Stokes, assuming the sphericity of particles:

$$F_S = 6\pi\mu Rv, \tag{6.4.4}$$

where μ is viscosity of a fluid, R is the particle's radius. However, in the process of motion of the small particles in gases are essentially violated those hydrodynamical conditions, by which the above formula was derived. The first scientist who brought his attention to this fact was Smoluchowski, who conducted the crude estimations of the coefficient K for the case of large lengths of the free pass of molecules λ. The main work on the refinement of the coefficient K for the case of medium's rarefaction is, undoubtedly, the experimental work by Millican. For the interpolation of his experimental data Millican gave the semi-empirical formula:

$$F = F_S \left(1 + A\frac{\lambda}{R}\right)^{-1}, \tag{6.4.5}$$

for which he himself stated the limit of applicability $\lambda/R \leq 0.5$. For the cases of larger relations λ/R one should use the formula of the type

$$F = F_S \left((B + A)\frac{\lambda}{R}\right)^{-1}. \tag{6.4.6}$$

The coefficients A and $A + B$ were determined by Millican experimentally. In spite of a careful conduction of the experiments, without the exact theoretical justification the indicated above dependence of the drag force on rarefaction of a medium did not lend a proper credibility even from Millican himself, and in his experiments on checking the theory of Brownian motion he gets rid of "the most inconvenient and inexact factor, connected with Brownian motion" by way of compensation of the resistance of the moving particles by some electrostatical force. The theoretical work by Einstein, made in 1924 by request of Millican, presents the approximate estimation of the sphere's drag: the exact value of that is obtained only in the free-molecular limit. The exact theoretical analysis might be realized only by way of solution of the Boltzmann equation, and the appearance of the works on that subject has become possible almost half a century later due to the development of electronic computers. The first computations in this area were limited either by the crude approximation of the distribution function, or by several approximations for deflection from the free-molecular limit. As it was done already earlier, one should specially mention the work by Cercignani and his colleagues, who obtained the results of the computation of the sphere's drag in the whole diapason of rarefaction, from the regime of continuous medium to the free-molecular flow.

The part of a characteristical parameter of the degree of rarefaction is played by the Knudsen number,

$$Kn = \frac{\lambda}{R}. \tag{6.4.7}$$

For the case of $Kn \to 0$ one obtains the value of drag by Stokes, while by $Kn \to \infty$ the curve is tending to the free-molecular regime with the value of drag obtained by Einstein. However, Einstein solved not the Boltzmann equation itself, but its approximate model, in which the integral of collisions is changed for the difference between the distribution function we are looking for and the equilibrium one. The alternation of collisional integral is explained by its tremendous complexity which introduces essential difficulties in the realization of computations. Presently, only the Monte Carlo method permitted to overcome difficulties of the computation of collisional integral and to obtain the solution of Boltzmann equation for the determination of the drag of a sphere in the regime of a transitional flow (see Khlopkov[104]). The dependence of a drag force on the number Kn was obtained in the reference cited and presented in Fig. 6.4.

The results presented above permit to obtain the theoretical values of the coefficient of diffusion and of the mean shifting within the whole range of the flow regimes from that of continuous medium and till the free-molecular one.

It is necessary to note the following fact. Millican was conducting the experiments on rendering more accurate results concerning the Stokes' law. When working on that with oil drops, he discovered an interesting kind of dependence: 89.5% of the molecules falling in the oil surface are reflected diffusely, and 10.5% of them are reflected specularly. If one builds up the diagram of dependence of the coefficient of diffusion and of the mean shifting on the number Kn just with the indicated correlation between the numbers of molecules having diffuse and specular reflection, then the coincidence of experimental data with the exact numerical solution of Boltzmann equation occurs to be within the limits of errors committed by experiment and by the conduction of computations — see Fig. 6.5.

Here D_0 and λ_{x_0} are coefficient of diffusion and mean value of shifting, which correspond to the classical solution by Stokes. As it is seen from diagrams, both diffusion and mean shifting are varied by the variation of the degree of rarefaction of a medium. The difference in values of the coefficient of diffusion and of the mean shifting, obtained by the Stokes' formula and by way of solution of Boltzmann equation, might achieve the several orders of a numerical value.

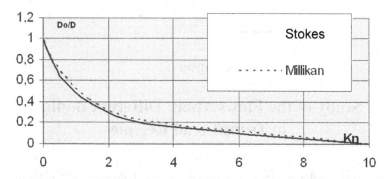

Fig. 6.5   Coefficient of diffusion of the Brownian particle in rarefied gas.

One should make a reservation in view of the fact, that the dependence of λ_S on *Kn* is valid only for those values of *t*, for which is valid formula (6.4.2). At the large values of *Kn* the interval of duration of free path might become comparable with the value of *t*, and that violates the main assumption, on the observation of which was based on the derivation of the formula mentioned.

Chapter 7

Study of the Flows About Different Bodies in Transitional Regime

The extremely violent development of the space technique and the high-altitude hypersonic aviation demands the possessment of reliable data on the aerodynamical characteristics in the wide range of flow regimes, beginning with that for continuous medium and ending with a free-molecular one. Considering the motion of the apparatuses in the lower layers of the atmosphere one usually comes to such problem, which might be solved within the frames of the theory of continuous medium or, more exactly, with the application of the equations by Navier–Stokes and by Euler. In their essence, these problems appear as the problems of the ordinary gas dynamics. In the other limiting case — that of the free-molecular flow, the integral of collision in Boltzmann equation turns into zero. In this case, the general solution of equation represents in itself the boundary function of distribution, which is conserved along the trajectories of particles. For the simple bodies the aerodynamical characteristics are obtained in the analytical form. The computation of the characteristics of flows and the summary of results is given in the monograph by Kogan.[8] The solution of the problems of flow about the bodies of complicated configuration for the cases of a free-molecular flow and of a hypersonic flow, which is nearly free-molecular, was executed by Perepukhov[52] with the help of Monte Carlo method and within the frame of a theory of the first collisions.

For the treatment of problems in the transitional regime, it would be necessary to solve the kinetic equation. Considered in the present chapter is the flow of a rarefied gas about bodies, at various numbers Kn and using the methodics, developed in Chaps. 3 and 4.

The setting of a problem of aerodynamical flow about body is executed in the following way. The field of flow in vicinity of the body within that flow is bounded by the area of the size of several lengths of the mean free path of molecules. This area is divided in cells, the linear size of which should not exceed the local length of a free path. At the frontal boundary of that area is set a condition of the absence of a gradient of distribution function. At the surface of a body within the flow is prescribed the law of interaction of the molecules and the surface. In the majority

114

of cases under consideration is set a condition of the completely diffuse reflection. The characteristical dimensionless parameters of the problem are $Re_0 = \frac{\rho_\infty V_\infty L}{\mu(T_0)}$, where Re_0 is the number by Reynolds, computed on the basis of the stagnation temperature, the analogue of Mach number, the number $\vec{S}_\infty = \frac{\vec{V}_\infty}{\sqrt{2RT_\infty}}$, and the temperature ratio $f_w = \frac{T_w}{T_0}$, where T_w is the temperature of a surface (see Gusev, Kogan, Perepukhov[108]).

Usually as the parameters of the problem of a flow about a body are taken also the angle of attack α (the angle between the direction of a vector $\vec{S}_\infty$ and the longitudinal axis of a body) and the species of a gas, which is characterized by the kind of dependence of the viscosity of gas on the temperature, $\mu \sim T^\omega$.

As the parameters of the results of the problem's solution should be taken the aerodynamical characteristics of the body in flow; the forces, the moments, the heat flow, acting on the body; the gasdynamical parameters in a field of flow.

7.1   Flows About the Planar Bodies

By way of using the kinetic equation by Krook and the Monte Carlo method calculation is made for the flow about the infinite wedge and about a cylinder of the rarefied gas in hypersonic regime. Calculated are the coefficients of drag and parameters of the field of flow for the wedge with various semi-angles by a vertex. The results of calculation are compared with those of other authors and with experimental data.

It is considered that the motion of a gas is described by the approximate kinetic equation — the equation by Krook:

$$\frac{df}{dt} = \nu(f_0 - f), \tag{7.1.1}$$

where f is the distribution function and ν is the frequency of collisions.

As it is assumed in Eq. (7.1.1), the distribution function after the collision appears as the most probable by the prescribed numbers of the colliding particles, their momenta and energies. By the direct modeling of the process with the help of Monte Carlo method the above assumption permits to work without the memorizing of the distribution function. The roaming of the trial molecule is realized according to the following scheme:

1. The molecule flies in from the boundary having the distribution function corresponding to the boundary conditions at infinity.
2. The length of a free path is determined in accordance with the frequency ν.
3. The velocity after collision is determined through the equilibrium function f_0.

4. By the interaction with the body's surface are observed the conditions of the nonflow-through with the prescribed law of molecular reflection, and so on, returning to the point 2 and proceeding until the molecule will leave the area of flow.

During the time of roaming computed within each cell is the new field of macroparameters, and after its stabilization (of the order of 10^3–10^4 trajectories) the roaming of particles is realized on the new field. The computation is executed by the method of successive approximations, until the results of the present approximation will not differ of those of the subsequent one with the prescribed accuracy.

7.1.1 *Flow about the wedge*

The computation of flow was executed by the various numbers S_∞, Re_∞, and α, where α is the semi-angle at the wedge's vertex. Using the results of computation (see Vlasov, Khlopkov[55–57,64]) and of the experiment,[108] one might come to the conclusion that in upstream direction the perturbations are transferred to the distance not exceeding 4–5 λ_∞. For that reason, the frontal boundary of flow was set at the distance 4–5 λ_∞. The boundaries above the body and behind of it were set at the distance several times larger. The step of a rectilinear network in X and Y directions in the field of flow was chosen to be constant. In the first approximation prescribed within the cells was either the free-molecular distribution of parameters, or the distribution, constant in the direction of flow, of the parameters equal to those at infinity. For the stabilization of flow with an accuracy up to 5% it would be sufficiently to consider about 10^4 in each iteration, while for the stabilization of the solution would be sufficiently to execute several iterations.

In view of the fact that by the moment of execution of the computations[64] were not yet known the published results, either numerical or experimental, on the flow about a wedge in the regimes studied in Ref. 64, the comparison of the results obtained was made with those, obtained for a flow about the plate under the angle of attack. Shown in Fig. 7.1 are the results of the present computation in their comparison with theoretical and experimental results on the flow about a plate under the angle of attack. The comparison conducted permits to come to the conclusion that at the large speeds and by the numbers Re_0 of the order of unity the aerodynamical characteristics for a wedge prove to be rather close to those for a plate.

Thus, the results of the computations conducted show that by the large supersonic speeds the qualitative picture of a flow about the wedge in a transitional regime corresponds to the physical picture of a flow about the plate under the angle

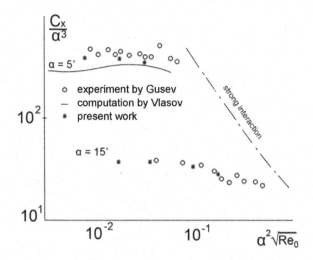

Fig. 7.1   Flow about a wedge.

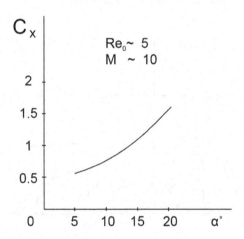

Fig. 7.2   Dependence of the drag of wedge on the half of angle by the apex.

of attack, which was very carefully studied both by theoretical way and by the experimental one (Figs. 7.2 and 7.3).

7.1.2   *Flow about the cylinder*

The computation of a flow was executed by the various parameters of the problem. As an example, presented in Fig. 7.4 is the field of density in its relation to the density at infinity ($S_\infty = 1$).

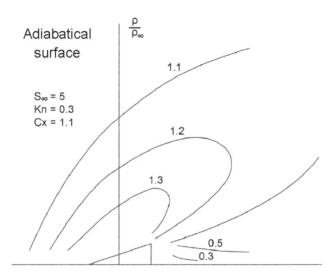

Fig. 7.3   The field of flow in vicinity of wedge.

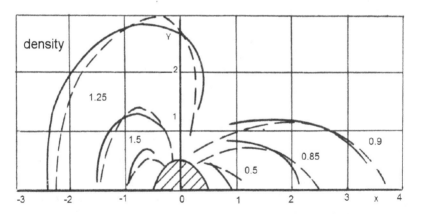

Fig. 7.4   The flow about a cylinder.

The dotted lines here and in some further figures correspond to the experiment.[107] The computation was executed also for the cylinder with the adiabatical surface, at $S = 5$. Presented in Fig. 7.5 is the field of flow above the cylinder at $Kn = 1$. The visibly small difference in the fields of flow, as compared with the results by Vogenitz and Bird (see Ref. 32), might be, apparently, explained by the different kinds of the initially used equations (in Ref. 32 with the help of Monte Carlo method is solved the complete Boltzmann equation).

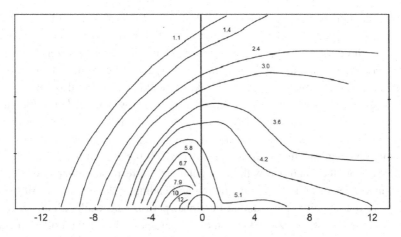

Fig. 7.5   The field of flow in vicinity of cylinder.

7.2   Flows About Axisymmetrical Bodies

The widespread interest to the problems of a flow about the axisymmetrical body in the transitional regime is explained, on the one hand, by the steadily extending front of the studies in the area of high-altitude, hypersonic aerodynamics, where the axisymmetrical flow might be consisted as the first approximation for the computation of a flow about the real apparatus. From the other hand, this is connected with a study of motion within the gaseous flow of the multitude of small particles with various physical processes occurring at the surface. By the experimental determination of the aerodynamical characteristics of bodies at large speeds of flow, at the present time the natural conditions cannot be reproduced in sufficient completeness at the existing experimental devices. In this connection, the special importance is attached to a such methodics of computation of the aerodynamical characteristics, with the help of which might be revealed the degree of influence of various parameters of a flow about a body. As it is known, the transitional regime is characterized by the fact, that the influence of viscosity and conductivity of heat is revealed not only within the thin layers at the walls and within the shock waves, as it happens by the usual flows about bodies, but also the influence spreads to the whole zone of flow. Therefore, for the description of such a flow it would be necessary to use the complete set of the Navier–Stokes equations ($N - S$) for the compressible fluid in the whole area of flow. As the degree of rarefaction of the medium is increased, it becomes to be necessary to consider gas as a totality of the large number of molecules, of which the behavior is described by the kinetic equation by Boltzmann (B).

The formulation of the problems of flows about axisymmetrical bodies possesses certain peculiarities. Generally speaking, the field of such a flow presents in itself a three-dimensional area, within which are built up the trajectories of particles. Having in mind to use the symmetry in respect to the azimuthal angle, two methods are used. In the first of these[39] "take of" within the field of flow introduced is the cylindrical coordinate system, and the field of a flow is divided in cells on the azimuthal angle (see Fig. 7.6). Taking into account the fact, that within all these cells the flow is just the same, the trajectories of trial particles are built only for one of cells. It means that from the cylinder which presents in itself the field of flow, is cut out the small part with the angular size corresponding to that of a cell. The trajectories of particles are beginning and ending at the boundaries of flow, that is, at the lateral surfaces and at the upper part of cell, while at the lacets is observed the law of specular reflection of the particles. This method of treatment is not free of the necessity of the upbuilding of three-dimensional trajectory, but, nevertheless, the necessary memory of a computer is not increased.

The second method of treatment is reduced to the use of a presentation of the kinetic equation in the cylindrical coordinates. Then the differential operator acquires the form

$$\frac{df}{dt} = \frac{\delta f}{\delta t} + \xi_r \frac{\delta f}{\delta r} + \xi_x \frac{\delta f}{\delta x} + \frac{\xi^2 \phi}{r} \frac{\delta f}{\delta \xi_r} - \frac{\xi_r \xi_\phi}{r} \frac{\delta f}{\delta \xi_\phi}.$$

This operator differs from the operator presented in Cartesian coordinate system by the presence of terms denoting the inertia forces, and, consequently, the rectilinear

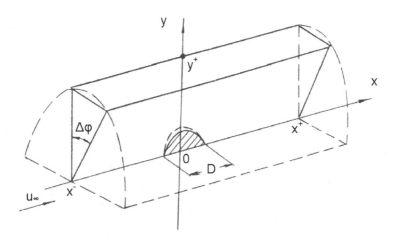

Fig. 7.6   The computational field for axisymmetrical flow about a body.

trajectories of particles are bended due to the forces, proportional to

$$\frac{\xi_\phi^2}{r} \quad \text{and} \quad -\frac{\xi_r \xi_\phi}{r}.$$

By the solution of problems of the flows about axisymmetrical bodies (a sphere, a cone), using the relaxational equation (7.1.1) in the form of model by Krook, it is usually sufficient to execute 5–7 iterations. In each of these iterations are taken 30–50 thousands trajectories, the size of cells is equal to 0.1–0.3 part of the length of free path of the molecules at infinity, and the error of the calculation of the integral characteristics oscillates around several percent, while the accuracy of computation of the field of flow, especially near the axis line, is essentially lower, and the error might reach the figures of 10–15%.

7.2.1 *Flow about the sphere*

From the practical point of view, the most interesting subject is presented by the behavior of the drag of a sphere within the whole range of the transitional regime of the flow about it. The computation of flow was carried out for a monatomic gas, for the Mach number M belonging to the range of values, corresponding to the supersonic and hypersonic flows. The Reynolds number Re_0 was varied from zero to several dozens. The value of temperature ratio t_w corresponded to the external cases of a cold ($t_w \approx 10^{-2}$) and of a hot ($t_w = 1$) body, the values of ω were taken as 1 and 1/2. As it is known, the Maxwellian molecules and the elastic spheres present in themselves two extremal expressions of the most soft and most rigid potentials of intermolecular interaction. It would be quite natural to assume that the values of aerodynamical characteristics for the mean values of parameters will appear between those for the aforementioned cases, though, generally speaking, this situation is not evident one.

Presented in Fig. 7.7 is the summary if theoretical and experimental results taken from Ref. 68, which are illustrating the influence of the above-mentioned parameters of flow on the behavior of the coefficient of drag of the sphere considered. These results are presented in the form of dependence of the coefficient of sphere's drag C_x on the temperature ratio t_w and within the whole range of Re_0 numbers. The computations corresponded to a solution of the kinetic equation by $t_w = 1$ (curve 1), $t_w = 0.1$ (curve 2), to the solution of Navier–Stokes equations with $t_w = 1$ (curve 7), and to the value obtained with the help of method by Newton (symbol Nw). The experimental data of various authors are presented for $t_w = 1$ (ordinary vertical lines) and for $t_w = 0.01$ (double vertical lines) and give a confirmation of the fact that $C_x(Re_0, t_w \sim 1)$ is larger than $C_x(Re_0, t_w \sim 0.01)$ within the whole range of Re_0 numbers. The qualitative influence of the condensation and

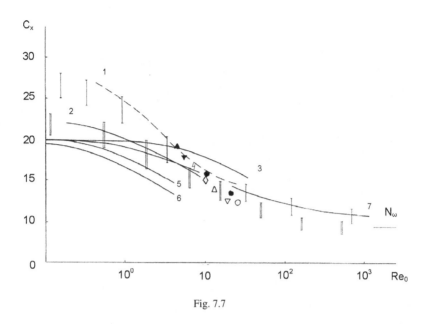

Fig. 7.7

evaporation from the sphere's surface by the gaseous flow about it is illustrated by the calculated curves of numbers 3, 4, 5, and 6. As it was noted earlier, the computation was carried out for the case of a uniformly evaporating (condensing) sphere. The curve 4 is approximately corresponding to the conditions of nonflow-through, the curve 3 — to the strong condensation at the surface, and the curves 5 and 6 — to the strong evaporation.

It is seen that the most expressive influence on aerodynamical characteristics of the body in flow, in the transitional regime, is made by the number Re_0. Evidently, the maximal variation occurs in vicinity of those values of Re_0, for which the description of flow should be carried out on the kinetical level. In all likelihood, such a behavior of C_x is explained by the fact that just within that range ($Re_0 \approx 1$) occurs the sharp change of the flow's structure connected with beginning of a formation of the shock wave. This process ends in the regime of viscous shock layer, and later on, as far as the thickness of a wave is diminishing, the flow stabilizes and by $M \gg 1$ the drag tends to its extremal value, obtained by the Newton's method. As it is known, the hypersonic stabilization is achieved by $M \geq 5$, and, therefore, the influence of Mach number on the aerodynamical characteristics will be revealed in the most pronounced degree at the comparatively small supersonic speeds. Thus, with the decrease of Mach number by $M < 5$ the value of C_x is increasing, and rather essentially. The increase of C_x by the small values of Re_0 is explained by the quality of contribution into the drag's value of the thermal dispersion of molecules, which is negligibly small by large M. The very important part in the transitional regime is played by the friction. By the approach to the nonviscous limit, the contribution of friction diminishes and the force of drag is determined, mainly, by the pressure.

The essential influence on behavior of C_x is made, also, by the value of temperature factor, and, just as it was in the preceding case, this influence is increased as soon as grows the degree of rarefaction of a medium, and reaches its maximum in the free-molecular case. At the large supersonic speeds the influence of a temperature ratio becomes, in comparison with that of the other parameters, to be the governing factor, and the difference in values reaches the figure of 20–25%. The explanation of such a behavior of C_x might be the following: in the free-molecular case the superiority of the drag of a hot body over that of a cold one is explained exclusively by the reactive force of the hot particles reflected from the body. As soon as the degree of rarefaction is diminished, and as the intermolecular collisions begin to play the essential part, the increased density ahead of the cold body makes the transfer of momentum of the oncoming flow to the surface to be more difficult. So far as the density of flow is increased, the influence of boundary conditions becomes less pronounced, and independently of these conditions the value of C_x is tending to be closer to its extremal figure.

Finally, in connection of the influence of the species of gas on the aerodynamical characteristics it would be possible to speak out in the following way. In the majority of cases considered, at the equal conditions the values of C_x for $\omega = 1$ appear to be somewhat lower than the same values for $\omega = 1/2$. In some cases this distinction might reach the figure of 10%, but, nevertheless, this situation deserves the further study.

The results of computations presented above do not reveal any exceeding of the coefficient of drag over its free-molecular value. This is true for any chosen values of parameters, though some time ago the possibility of such an exceeding was actively discussed. The qualitative comparison with experimental data leads, basically, to the confirmation of conclusions of the analysis conducted.

7.2.2   *Flow about the cone*

For the computation of aerodynamical characteristics of a cone in the transitional regime was used the method developed in Ref. 36. Computations of the flow about a cone on the basis of kinetic equations were carried out in the series of works. Here we are following the studies[67,69] devoted to the flow about a cone, where the computation is carried out by the various values of the azimuthal semi-angles, various Reynolds numbers, and various values of the temperature ratio.

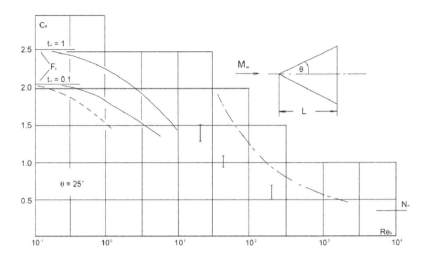

Fig. 7.8   Dependence of the cone's drag on various parameters of flow.

Presented in Fig. 7.8 is the dependence of the coefficient of drag of a cone
on the various parameters of the oncoming flow. Just as it was in the preceding case, the strongest influence on the variations of drag was rendered by the
Reynolds number. The results of computations executed in the cited references
(the solid lines at the Figures) are presented for the numbers Re_0 beginning from
zero and up to several dozens. The general picture of the behavior of C_x might be
demonstrated by way of introduction into the graphs the values of drag obtained
through the use of a theory of the first collisions with the help of a variational
principle. The curves are corresponding to the theoretical data of the viscous interaction, while the experimental data are denoted by the vertical lines. Nonviscous
regime ($Re_0 \to \infty$) and free-molecular one ($Re_0 \to 0$) are denoted by the straight
horizontal lines.

The strong influence on the behavior of C_x is rendered by the geometry of a
cone. By the large values of the semi-angles by the vertex ($\theta \geq 20°$) the behavior
of C_x is qualitatively similar to that of the drag of a blunted body, for example, of
a sphere. The value of C_x is monotonously diminishing with the increase of Re_0
from the free-molecular value and till the nonviscous limit. And just as it happens
for a blunted body, the regime of hypersonic stabilization sets in by $M \geq 5$.

The next factor, influencing the value of drag at large θ, is the temperature ratio,
and once again it is evident that the influence of this ratio proves to be maximal in
the free-molecular area. It is necessary to note that by the diminishment of the semi-
angle by the vertex the behavior of drag becomes more and more different from

the behavior of C_x for the blunted body. Thus, for example, this behavior becomes less and less sensitive in respect to the temperature ratio (see Fig. 7.8, where by the points are denoted the results of numerical solution of the Boltzmann equation), and by the values $\theta < 10°$ the character of this behavior is changed qualitatively. The clearly visible maximum of the function $C_x(Re_0)$ is revealed in vicinity of $Re_0 \approx 1$, and the increase of C_x proves to be more essential for the smaller values of the semi-angle by the vertex and for the smaller temperature ratios.

7.3 Influence of the Evaporation (Condensation) on the Aerodynamical Resistance of a Sphere by the Supersonic Flow About It

In connection with the development of the rocket-spatial techniques lately the more and more importance acquire the problems connected with a motion within the gaseous medium, by the small Reynolds numbers, of the bodies subject to the sublimation and condensation. Moreover, the similar problems are met in meteorology, in the vacuum and space technologies, and so on. As one of the most important characteristics by the studies of multiphase systems appears the aerodynamical loading, acting on the sublimating or condensating bodies (particles) from the part of a gaseous flow.

In the works by Nikolskii and Khlopkov (see Refs. 109–111) studied with the help of theoretical computations and experimental methods is the influence of the evaporation and condensation from the sphere's surface on its aerodynamical resistance. The computation was carried out by the method of direct statistical modeling (see Khlopkov[39]). Solved is the model problem of the supersonic flow about a sphere, with the variable degree of evaporation or condensation from the surface, in the dependence on the degree of rarefaction of a medium. The experiments were conducted by Ju. V. Nikolskii in the vacuum wind tunnel with the Mach number of the aerial flow $M = 7.05$, at the stagnation pressure $P_0 = 10$ torr and at the various stagnation temperatures: $T_0 = 294$ K; $T_0 = 393$ K; $T_0 = 493$ K; $T_0 = 572$ K. The measurement of the force of resistance of the sphere was carried out with the help of aerodynamical scale which provides the error of measurement within the limits of $\pm 0.2\%$.

Considered by the computation is the steady supersonic flow of rarefied gas about the axisymmetrical body. The field of a flow in vicinity of the body in it is bounded by a finite area, at the frontal boundary of which is prescribed the equilibrium distribution function:

$$f_\infty = \frac{\rho_\infty}{m}(2\pi RT_\infty)^{-3/2} \exp(-(\xi - u_\infty)^2/2RT_\infty)).$$

At the side and background surfaces of the area is set the boundary condition of the absence of a gradient of the distribution function. At the surface of the body in flow is prescribed the law of interaction of molecules with a surface. Thus, for the case of nonflow-through the reflection of molecules from the surface is considered to be the diffuse one, with a temperature, equal to that of a surface of the wall:

$$f_w = \frac{\rho_w}{m}(2\pi RT_w)^{-3/2}\exp(-\xi^2/2RT_w),$$

where ρ_w and T_w are the density and the temperature of the molecules reflected from the surface; the value ρ_w is determined from the nonflow-through condition. In the case, when modeled are the conditions of evaporation or condensation at the surface, it is assumed that the molecules of the oncoming flow are completely captured by the body, while from the body's surface is realized the uniform evaporation, having the distribution function with the prescribed values ρ_w and T_w.

The models of spheres for the experiment were produced from the crystallic camphor. The prescribed quantity of the camphor powder was pressed in spherical form in such a way that the density of a substance of these models was constant, $\gamma = 0.99$. The choice of the camphor as a material for models was stipulated by the considerable speed of sublimation in the conditions of vacuum. In the process of experiment the model was photographed. Its form at the different moments of time is presented in Fig. 7.9 with two values of the stagnation temperature, $T_0 = 294$ K and $T_0 = 493$ K.

The value of a unit mass flow of a camphor G from the model's surface was determined experimentally, by the variation of the model's volume during the unit

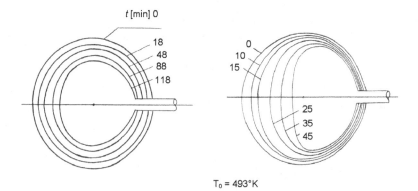

Fig. 7.9   The form of evaporating sphere at different time moments.

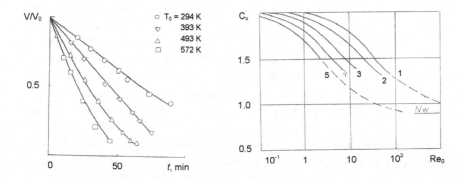

Fig. 7.10   The mass expenditure of camphor from the model's surface and the sphere's drag.

time. The values of G in relation to $\rho_0 u_0$ of the gaseous flow, at the various values of the stagnation temperature are presented. Here one should note that with the increase of a temperature of the gaseous flow becomes to be more pronounced the nonuniformity of the evaporation of substance from the surface of a model (see Fig. 7.10).

The drag force of the evaporating model was measured at the initial moment of time ($t \sim 5\,\text{min}$), when the model's form was nearly spherical. As the initial temperature of the model's surface was taken the room temperature, $T_W = 294\,\text{K}$.

Presented in Fig. 7.7 are the computational and experimental data for the flow about a sphere by the various parameters of the oncoming flow and by the various conditions of the interaction of a gaseous flow with the sphere's surface. The results are presented in the form of dependence of the coefficient of drag of the sphere C_x on the parameter of a degree of rarefaction of medium Re_0. Denoted by the lines are computational data, while denoted by vertical lines and points of various configuration are experimental data. For the convenience of reading the conventional designations are listed. The dependence of behavior of C_x on the temperature ratio within the whole range of the numbers Re_0 is presented by the results of numerical solution of the kinetic equation by $t_W = 1$ (curve 1), $t_W = 0.1$ (curve 2), solution of Navier–Stokes equations by $t_W = 1$ (curve 7), and by the value, obtained with the help of Newton's method. The experimental data of various authors, gathered in the preceding paragraph by $t_W = 1$ (ordinary vertical lines) and by $t_W = 0.01$ (double vertical lines) confirm the dependence

$$C_x(Re_0, t_W \sim 1) > C_x(Re_0, t_W \sim 0.01),$$

within the whole range of numbers Re_0. The qualitative influence of the condensation and evaporation from the sphere's surface within the gaseous flow is illustrated by the computational curves 3, 4, 5, and 6. As it was already noted, the computations were carried out for the uniformly evaporating (condensating) sphere. The curve 4 is approximately corresponding to the conditions of nonflow-through, the curve 3 — to the strong condensation at the surface, and curves 5 and 6 — to the strong evaporation. The computational results indicate at the strong dependence of a diminishment of C_x on the intensity of evaporation. One might assume that the diminishment of C_x in the presence of evaporation will be observed in the experiment, as well. Presented at the same figure are also the experimental data on the flow about a sphere, obtained in the work referred. The solid points correspond to the results of measurement of the force of resistance of metallical sphere in the flow obeying the conditions of nonflow-through at the surface, coinciding with the summarized data referred in the preceding paragraph. By the small circles are designed the results of measurement of the resistance of a sphere produced of the camphor, with the conditions of a weak evaporation from the surface. It is to be noted that the presented here comparison of the computational and experimental data on the resistance of a sublimating sphere has a qualitative character. The conditions of experiment correspond to the two-component model of a flow about an evaporating sphere (oncoming flow — the air, evaporating substance — the camphor), while the computations were carried out for a flow about the sphere evaporating gas of the same type as in the oncoming flow. Nevertheless, one might state that the evaporation from the surface of a sphere in the supersonic flow leads to a diminishment of its aerodynamical resistance.

7.4 Computation of the Steady Regime of a Flow About a Body and of the Profile Resistance in a Viscous Gas (See A.S. Petrov)

As it was already noted earlier, in the overwhelming majority of the cases considered, the specialists in the applied aerodynamics were interested just in stationary or steady in time aerodynamical forces. If one limits himself with a consideration of the class of well-streamlined bodies of the type of a wing profile, and excluded from that the bodies wittingly not being subject to the stationary flow about them, like a plate posed perpendicularly to a flow, then, apparently, the steady regimes of the flows about bodies of such a type are always existing. In this case, all the stationary aerodynamical characteristics might be found as the corresponding limits of the nonstationary ones at $t \rightarrow \infty$.

From the positions of the method presented above the existence of a steady in time regime of a flow about body means the transformation of an accidental process

(5.6.6), which plays, ultimately, the decisive role in the problem's solution, into the so-called *stationary* accidental process, during the development of which both mathematical expectations and the dispersions of process cease their variations from one temporal step to another.

Then it becomes evident that the "gambling" of the process (5.6.6), carried out N_0 times at each temporal step, is equivalent to a single "gambling" at temporal steps of N_0 number. In this connection arises the possibility to "gamble" the process just once at each temporal step and to find the vorticity in the cell, which is afterwards averaged over N_0 temporal steps. The existence of the averaged in time characteristics of the stationary accidental processes, with the unit probability, appears as a strict sequence of the ergodical theory by Birkhoff–Khinchin. And now, after limiting ourselves by a consideration of the class of bodies, for which the steady regime of flow about them exists, we see that the problem is essentially simplified. First, there is no necessity to divide all the space of flow into the finite elements. Second, arises the possibility to follow the individual motion of all the elements of the zone of vorticity, and this possibility leads to maximal approach of the algorithm to be proposed to that of the method of discrete vortices, which is widely used for the ideal fluid. The division of space into the finite elements is kept only in those areas where it is necessary to calculate the values of vorticity, for example, in the wake or in the area near a wall, for the calculation of the frictional resistance.

The value of vorticity is calculated in the following way. Let the cell have the area ΔS_j and coordinate $\vec{q}_j$. At each of the moments of time calculated are the number of vortices within the cell and their total intensity $\Gamma_j(t)$:

$$\Gamma_j(t) = \sum_{i=1}^{N} \Gamma_i.$$

Here N is a number of vortices in the cell, Γ_i — their intensities.

Now the mean value of a vorticity at the moment of time t, which is averaged within the preceding interval of time $T = t - \tau$, will be determined by the expression

$$\bar{\Omega}(t, \vec{q}_j) = \frac{1}{\Delta S_j} \lim_{t \to \infty} \frac{1}{T} \int_{\tau}^{t} \Gamma_j(t) dt. \qquad (7.4.1)$$

The mean value of vorticity in the area near the wall might be used for the computation of the frictional resistance. The mean in time value of the longitudinal velocity, which is necessary, for example, for the calculation of the profile resistance of a body, is found by the usual way,

$$\bar{V}(t, \vec{q}) = \lim_{t \to \infty} \frac{1}{t - \tau} \int_{\tau}^{t} V_x(t, \vec{q}) dt.$$

7.4.1   *Flow about the plate*

The way to the solution of a problem, proposed here, proves to be most grounded one for the obtainment of the characteristics of the steady flow, which is typical for well-streamlined bodies. However, it does not mean that the same way cannot be used for computation of the nonstationary characteristics of the badly streamlined bodies. For such a computation it would be necessary that the time of averaging $T = t - \tau$ is far less than the characteristical time of the nonstationary aerodynamical process arising by the flow about body. This demand puts the more rigid limitations on the step in time,[135−137]

$$\Delta \bar{h} \ll \frac{1}{\sqrt{Re}}, \quad \Delta \bar{t} \approx \Delta \bar{h}, \quad N \gg \sqrt{Re}, \quad T \gg 1.$$

Shown in Fig. 7.11 are the structures of the vortical zones, appearing by the flow about a plate at the moment of time $T = 4$ from the time of a start. The plate is located at zero angle of attack along the axis OX, at the segment $(-1, 0)$.

It is seen that with the decrease of Reynolds number the vortical zone is "swelling", the thickness of a boundary layer is growing. This happens due to the fact that by the decrease of the Reynolds number the dispersion of a function

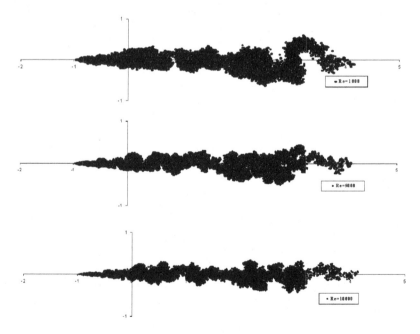

Fig. 7.11   Structure of the vortical zones behind a flat plate.

$f(\tau, \vec{p}_i, t, \vec{q}_j)$ is increased and, simultaneously, is increasing the amplitude of the accidental term in the expression (5.6.6).

The method proposed permits to carry out with satisfactory accuracy the computation of the steady profile of velocity within the boundary layer without the introduction of the special variables. Undoubtedly, such a computation, especially at the large Reynolds members, demands for the considerable computational resources and certain amount of time for averaging, if one wishes to obtain the steadification of the parameters to be found. Presented in Fig. 7.12 is the computational profile of the steadified velocity within the boundary layer of the plate and the comparison of that profile with a well-known exact solution.

By the name of reduced ordinate is understood the ratio of the ordinate to the thickness of a boundary layer on the plate: $\bar{y} = y\sqrt{Re}/5$.

Presence of the information on the velocity profile within the boundary layer is already permitting to carry out the computation of the frictional strain at the surface of a body in flow. In Fig. 7.13, the results of these computations are presented in comparison with the exact solution by Blasius for the boundary-layer equations.

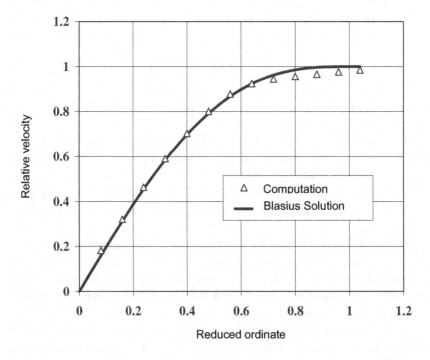

Fig. 7.12 Profile of the velocity in boundary layer.

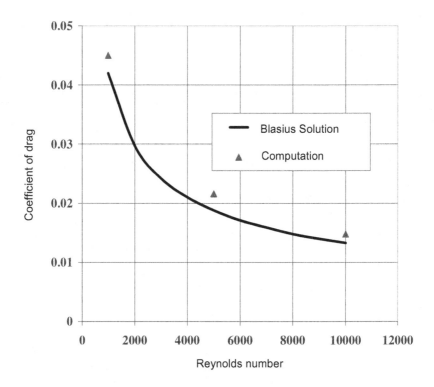

Fig. 7.13   Computation of a drag of a flat plate.

As a commentary it might be said that the reported above computations of the plate's drag give, in the comparison with a solution by Blasius, rather close, but stably overrisen values of that drag.

The method proposed permits to carry out, and sufficiently simply, the computations of the averaged values of parameters within the wake behind the body in flow.

Presented in Figs. 7.14 and 7.15 are the examples of the computational results concerning the longitudinal velocity of the base flow and of the averaged vorticity in the cross section of integration, located at the distance $\Delta \bar{x} = 0.4$ behind the trailing edge of the plate found in the longitudinal flow of a viscous, incompressible fluid, with the Reynolds number $Re = 1000$.

The distance from the wake's axis is given in the absolute units. Thickness of the boundary layer at the trailing edge of a plate by these conditions is equal to $\delta \approx 0.161$.

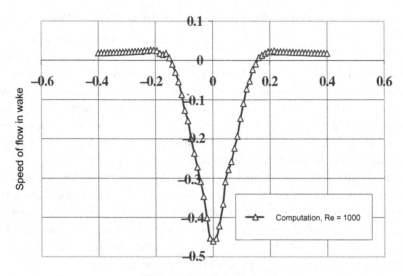

Fig. 7.14   Speed of the flow in wake.

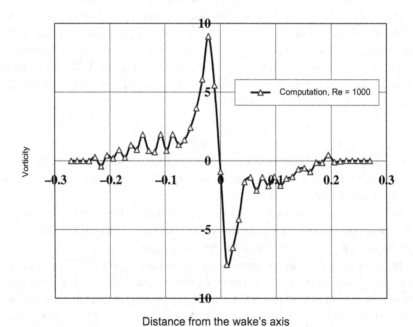

Fig. 7.15   Vorticity of flow in a wake behind a flat plate.

7.4.2 *Flow about the cylinder with the complicated boundary conditions*

Devoted to the flow of viscous fluid about a cylinder, studied by the method of statistical modeling, is the classical work by Chorin.[146] In the quality of a next illustration of the possibilities of the method, discussed earlier, the solution is built up for the flow of the incompressible viscous fluid about the circular cylinder, taking into account the suction of a boundary layer. The geometrical simplicity and the extremal complexity of the tearing-off processes arising in the flow about a body, make of the circular cylinder a classical "touchstone" for all the methods of solution of the problems involving Navier–Stokes equations.

Among all the methods of control of a boundary layer, the suction of that layer through the porous (perforated) surface of a body, or through the crack of a special construction, appears to be the most usable and practically important. The majority of theoretical methods is based on the boundary-layer theory, and, consequently, the rate of a suction or of a blow-away, should be sufficiently small for leaving out of the influence the external potential flow about a body,

$$V_w \leq \frac{V_\infty}{\sqrt{Re}}. \tag{7.4.2}$$

As one of the most interesting problems in that area appears that of a determination of such a rate of suction of the boundary layer, by which is completely ceased the tearing of flow off the body's surface.[147] The characteristical feature of the problems of that type proves to be the extremal value of the rate of suction, because its further increase is simply senseless. The value of that rate for the badly streamlined bodies might be rather large. For example, in the case of a circular cylinder this value, according to Prandtl, is equal to

$$V_w = \frac{4.35 V_\infty}{\sqrt{Re}}. \tag{7.4.3}$$

Judging by the order of its size, this value of a rate of suction obeys, formally, the limitation of (7.4.2), but, evidently, it lies sufficiently near to the limit of applicability of the boundary-layer theory.

In this connection, it might prove to be interesting to build up the solution of this problem with the help of Navier–Stokes equations, without introduction of limitations on the rate of suction of the boundary layer.

In the general case this problem is formulated in the following way. Let the planar solid body, bounded by the contour L, at the moment of time $t = 0$ is instantaneously set in motion with velocity V_w within the volume of viscous, incompressible fluid. Simultaneously, switched on is the suction of boundary layer, which is arbitrarily distributed over the surface of a body within the flow. The suction is carried out in direction of a normal to surface, and its intensity is not limited.

The problem of that type in the coordinate system, connected with a body, is reduced to a solution of the nonstationary equations by Navier–Stokes (Helmholtz):

$$\frac{\partial \Omega}{\partial t} + V_x \frac{\partial \Omega}{\partial x} + V_y \frac{\partial \Omega}{\partial y} = \nu \, \Delta \Omega,$$

$$\mathrm{div}\, \vec{V} = 0,$$

with the initial and boundary conditions

$$\bar{V}(t, \vec{q})_{t=0} = \vec{V}_0(\vec{q}), \quad V_\tau(t, \vec{q})_{q \in L} = 0, \quad V_n(t, \vec{q})_{q \in L} = V_w(t, \vec{q}).$$

Here $\vec{V}_0$ is the initial, potential field of velocities in flow, V_w is the rate of suction.

This problem was solved numerically by the method proposed, for the circular cylinder with uniformly distributed suction, in the range of Reynolds numbers $Re = 1000$–$10,000$ and for the rates of suction $V_w = (+0.02, -0.2)V_\infty$. Here and further on the condition $V_w < 0$ corresponds to the suction of boundary layer, and condition $V_w > 0$ corresponds to the blow-away.

Instead of the rate of suction or blow-away one usually introduces the coefficient of the voluminous expenditure, equal to

$$C_q = -\frac{V_w}{V_\infty}.$$

The range of its variation during the computations was $C_q = (-0.02, +0.2)$. It is necessary to note the maximal values of the coefficients of voluminous expenditure, studied in the problem stated, are considerably surpassing those values which are admissible in the theory of boundary layer by the corresponding Reynolds numbers. In the process of computations, by way of the averaging over sufficiently large interval of a time, obtained with the help of the expression (7.4.1) were the stable values of vorticity in the cells adjacent to a surface, and, after that, the local and the total values of C_f, the friction stress by the various values of the coefficient of voluminous expenditure C_q, and of Reynolds numbers. By the further intensification of a suction the coefficient of friction is not, practically, variating. It seems to be logical to connect this phenomenon with the steadification of the tearless regime of a flow and of the maximal, for practical applications, value of the coefficient of a voluminous expenditure. Evidently, the further intensification of suction would be senseless.

For the confirmation of that assumption analyzed were the distributions of local friction coefficient over the surface of a cylinder, in its dependence of the coefficient of expenditure. These dependences are presented in Fig. 7.17 by the Reynolds number $Re = 10,000$.

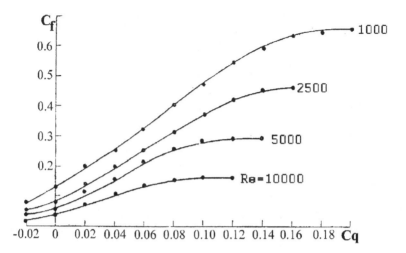

Fig. 7.16   Dependence of the total drag's coefficient on the volume expenditure.

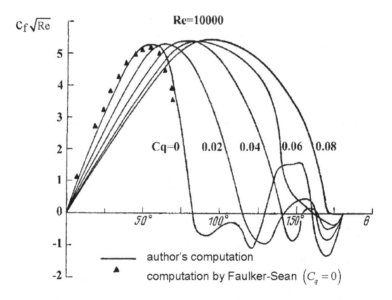

Fig. 7.17   Local coefficient of the friction by various intensities of the blow-out.

In the absence of the boundary layer's suction, up to the point of the first tear-off of the flow, the friction's distribution obtained coincides satisfactorily with the computational results according to the boundary-layer theory using the experimental distribution of the pressure.[148] Let us follow the transference of the first point of

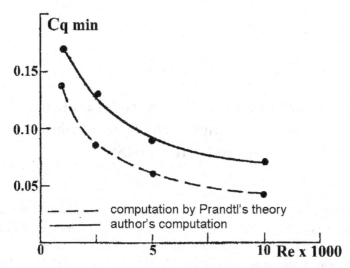

Fig. 7.18 The minimal value of a blow-out intensity eliminating the breakdown of boundary layer on cylinder.

the flow's tear-off in its dependence on the intensity of a suction. With the intensification of suction the point of a tear-off is regularly transferred downstream and by some value of suction occurs to be located at the lower critical point. This value of the intensity of suction is, practically, coinciding with the beginning of a smoothing of the curves of dependence in Fig. 7.16. Thus, one might state that by the Reynolds number $Re = 10,000$ the practically tearless flow about a circular cylinder is setting by the coefficient of voluminous expenditure for the suction of boundary layer equal to $C_q \approx 0.08$. Let us consider this value of C_q as minimally needed for the steadification of a tearless flow about a cylinder. In the similar way were found the minimally needed values of C_q for other Reynolds numbers. Presented in Fig. 7.18 is the dependence on Reynolds number of the minimal value of C_q, eliminating the tear-off. Presented in the same figure are the results of computation according to the formula by Prandtl, which was obtained with the help of the boundary-layer theory. Observed is the qualitative agreement of the results, although the values of C_q, obtained by the present computations, are somewhat higher.

It is possible that by the larger Reynolds numbers the coincidence between the results obtained and those of the boundary-layer theory will be better than here. However, like any other method based on the solution of the complete Navier–Stokes equations, for the achievement of high accuracy by the large Reynolds numbers the method proposed will demand for the considerable computational resources.

Chapter 8

Determination of the Aerodynamical Characteristics of the Returnable Space Systems (RSS)

The large amount of the aerodynamical computations and the diversity of schemes of the flying systems, which are characteristical for the project studies and for the first stages of project activity, are stipulating the necessity of development of the engineering methods of computation of the aerodynamical characteristics of the apparatus of the arbitrary form, by all the possible regimes of flight. There is a whole number of approaches to the construction of the engineering methodics of that type, based on the accessible theoretical and experimental information of the force loading at these regimes of a flight (see Barantsev, Galkin, Eropheev, Tolstykh, Bunimovich, Chistolinov, Bass, Shvedov, Eremejev, Khlopkov[84–96,101]).

8.1   Methodics of the Description of a Surface

One of the main aspects of the methodics of computation of the aerodynamical characteristics of the apparatus of the arbitrary configuration appears the choice of the way of a geometrical description of the surface. The methods of description of the complicated surfaces might be divided into two main groups: the mathematical approximation of a surface and the spatial distribution of the large number of points of this surface, with the help of which is reconstructed the system of elementary grounds. Usually to the main deficiencies of the first group of methods are brought the mathematical difficulties of approximation of the complicated and essentially nonlinear surfaces by means of a small number of the control points, whereas to the deficiencies of the second group — the difficulties in preparation of the starting data, which is connected with necessity of the taking from a draught of a large number of coordinates of the points of a surface. Presently, we are using a combination of these groups of methods: the second one is chosen due to the comparative simplicity and universality of prescription of the control points, and, ultimately, of the reconstruction of elementary grounds, while for the restoration of a surface through the control points the body modeled is divided into a series of specific parts (the wing, the nose part, the base part of a fuselage, etc.), and for each of these parts is carried out the quadratical interpolation over the control points.

The surface of the apparatus as a whole is divided into totality of the initial parts. For example, for the configuration presented in Fig. 8.1 (the model "Shuttle-Bouran") division is made into three parts: the fuselage, the keel, and the wing (due to the symmetry considered is only the right part of the draught).

In Fig. 8.2 (promising model of the hypersonic flying apparatus (HFA)), division is realized into two parts: the fuselage (pointed semicone) and wings. In Fig. 8.3 (complex "Bouran-Energia") — in six parts: three blunted cylinders (a rocket with accelerators) and three parts of "Bouran" properly. For each of the parts are introduced the axes (x, y', z'), which appears as those of cylindrical system of coordinates. The axes are divided into the finite number of characteristical

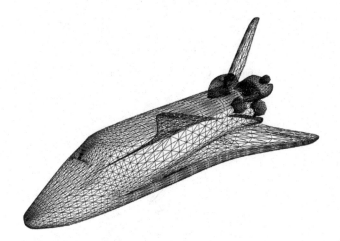

Fig. 8.1   Typical geometry of ASS.

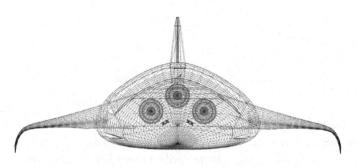

Fig. 8.2   Prospective hypersonic flying apparatus.

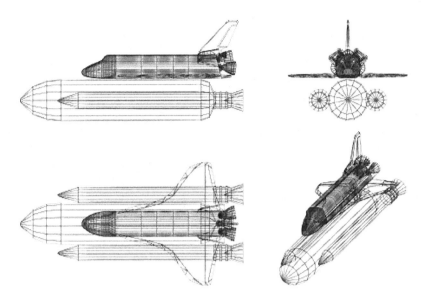

Fig. 8.3   Complex "Buran-Energia".

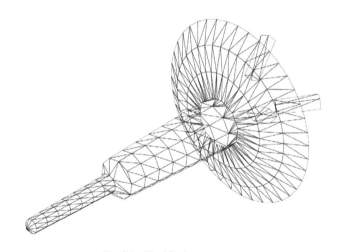

Fig. 8.4   The Marsian penetrator.

points given by the parameters x_i, y_i, z_i. In these points in the cylindrical system are prescribed the cross sections: φ_j; R_{ij}; φ_{yj}; R_{yij}; φ_{zj}; R_{zij}. Dependently on the form of a cross section, it might be prescribed either in a discrete form, or in an analytical one (Fig 8.4).

For the more accurate description of a surface in the intermediate points foreseen is the interpolational procedure. These intermediate points at the axes and values of the angles are interpolated according to the formulae of linear interpolation

$$x_i = \frac{1}{2}\left(x_{\frac{i-1}{2}} + x_{\frac{i+1}{2}}\right), \quad \varphi_i = \frac{1}{2}\left(\varphi_{\frac{i-1}{2}} + \varphi_{\frac{i+1}{2}}\right).$$

With the help of the interpolational multinominal by Lagrange the values of radii are interpolated twice — over φ and over x:

$$R(a) = \sum_{i=1}^{3} R(a_i) \prod_{j \neq 1} \frac{a - a_j}{a_i - a_j},$$

where symbols a correspond to the values of φ and x at the points of interpolation.

Thus, with the accuracy needed are prescribed the starting points at the surface. But there is yet a question of the way of stretching of the surface of apparatus on the framework constructed. As it was already noted, for the achievement of our aim will be suitable the linear approximation, and, therefore, as the main element will be considered the linear one, which presents in itself a triangle constructed on the three nearest points. The apices of these triangles in Cartesian coordinates are, for different parts, determined by the formulae:

for the fuselage

$$r = \begin{pmatrix} x \\ y \\ z \end{pmatrix} = \begin{pmatrix} x_i \\ R_{ij} \cos \varphi_j \\ R_{ij} \sin \varphi_j \end{pmatrix},$$

for the wing

$$\begin{pmatrix} x \\ y \\ z \end{pmatrix} = \begin{pmatrix} x_0 + z_i \cos \alpha_z - R_{zij} \cos \gamma_{zi} \\ y_0 + z_i \sin \alpha_z + R_{zij} \sin \varphi_{zj} \\ z_0 + z_i \cos \alpha_z \cos \beta_z - R_{zij} \cos \varphi_{zj} \sin \gamma_{zi} \end{pmatrix},$$

where (x_0, y_0, z_0) are the starting coordinates of the wing's axis z', α_z is the angle of inclination of the wing's axis in respect to a plane $y = 0$, β_{zi} is the angle of inclination of the wing's axis in respect to a z axis, γ_{zi} is the angle of inclination of the prescribed cross sections at the axis z'.

For the complete prescription of the element it is necessary to determine its orientation and the area of its surface. Let $a = r_2 - r_1$, $b = r_3 - r_1$ are the elements of the vector's formation. Then the area of element equals to

$$S = \frac{1}{2}(a \times b),$$

and the normal to its surface is

$$n = (a \times b)/(|a||b|).$$

The estimation of an error of the approximation by the linear elements, when the processing is realized at the free-molecular regime of a flow, leads to satisfactory results. Thus, for the approximation of a cone by the computation of its drag with accuracy of 5% (the mean value of error for the statistical methods) it is necessary to have, approximately, 10 elements, and for the approximation of a sphere — 100 elements. The one and unique application of the interpolational procedure decreases the error by one order.

8.2 Methodics of Calculation of the Aerodynamical Characteristics of the Flying Apparatus in the Conditions of a Free-Molecular Flow

The problem of determination of the aerodynamical characteristics is set in a usual way. Coming from the infinity to the body is the free-molecular flow of particles with an equilibrium distribution function

$$f\infty = n\infty \left(\frac{m}{2\pi k T\infty}\right)^{\frac{3}{2}} \exp\left[-\frac{m}{2kT\infty}(\xi - V)^2\right].$$

When considering the general case, it would be necessary to prescribe the connection between distribution functions of the falling molecules, $f\infty$, and of the reflected ones, f_r. Theoretically, such a connection is represented by the integral equation, holding on each element of the surface:

$$f_r = \int_{(\xi n) \leq 0} K(\xi, \xi_r) f\infty d\xi,$$

where the form of a kernel $K(\xi, \xi_r)$ depends on the physical and chemical properties of a surface and of a gas.

The existing theoretical and experimental data are not sufficient for the trustworthy judgment on the properties of the kernel. The experiment provides the most reliable results only for the formulation of a connection between the flow-like characteristics of the oncoming particles and the reflected ones through the coefficients of accommodation σ_n, σ_τ, and σ_e. If one would not consider the field of flow, then the values of accommodation coefficients are completely determinating the problem's formulation. The values of normal p_n and tangential p_τ momentum, as well as the value of heat flow q for the element of a surface, in the case of a hypersonic flow looks as

$$p_n = \frac{\rho V^2}{2} \left[(2 - \sigma_n)2 \sin^2 \theta + \sigma_n \sqrt{\frac{\chi - 1}{\chi}} t_w \pi \sin \theta \right],$$

$$p_\tau = \frac{\rho V^2}{2} \sigma_\tau 2 \sin \theta \cos \theta,$$

$$q = \frac{\rho V^2}{2} \sigma_l \left(1 - 2\frac{\chi - 1}{\chi} t_w \right) \sin \theta.$$

To check the capacity of work for the methodics, presented above, were carried out the systematical computations for whole set of bodies: semisphere, cone with the various semi-angles by a vertex, blunted semicones, and blunted semicones with wings. The computations were carried out with all angles of attack, from 0 to 180°, by the various parameters of the oncoming flow and various boundary conditions. For the bodies listed above one has, by certain values of parameters, both theoretical and experimental results.

Thus, using the methodics described and the prepared program of the computations, one would be able with a sufficient accuracy to calculate the aerodynamical characteristics for the wide variety of bodies, including the apparatus of complicated form, for which the interference between the various parts of these bodies does not play an essential role.

8.3 The Engineering Methodics of the Computation of Aerodynamical Characteristics of the Bodies of Complicated Form in a Transitional Regime (see Galkin, Eropheev, Tolstykh[85])

For development of the engineering methods of computation of the aerodynamical characteristics of the bodies of complicated form in a transitional regime of flight, beginning with the free-molecular flow and ending with that of a continuous medium, in the last years are widely used the approximate methods based on the hypothesis of locality.[84] In the present study we are using the approach, the essence of which is reduced to the assumption of an independence of the aerodynamical loads on the element of body's surface from the orientation of other elements. As the basis of the program of computation of the aerodynamical characteristics of the complicated bodies in transitional regime is accepted the method described in Sec. 8.2 for the free-molecular flows.

According to the hypothesis of locality it is assumed that the flow of momentum at the body's surface is determined by the local angle of a fall-down. Here we are using the following expressions for the elementary forces of a pressure and of a friction[85]:

$$p = p_0 \sin^2 \theta + p_1 \sin \theta,$$

$$\tau = \tau_0 \sin \theta \cos \theta.$$

The given above coefficients p_0, p_1, τ_0 (the coefficients of the regime of flow) depend on the Reynolds number $Re_0 = \rho_\infty u_\infty L / \mu_0$, the temperature ratio $t_w = T_w / T_0$, the coefficients of accommodation and the ratio of specific heats χ (L — the characteristical size, $\mu = \mu(T_0)$ — coefficient of viscosity, T_0 and T_w — stagnation temperature and temperature of a surface).

The independence of the coefficients of regime in the hypersonic case should provide the transition to the free-molecular values by $Re_0 \to 0$ and to the values of Newton's theory and of the methods of thin tangential wedges or cones by $Re_0 \to \infty$. Proposed on the basis of analysis of the computational and experimental data are the empirical formulae

$$p_0 = p_\infty + [p_\infty(2 - \alpha_n) - p_\infty]p_1/z,$$

$$p_1 = z \exp[-(0.125 + 0.078 t_w) Re_{00}\text{TMTM}],$$

$$\tau_0 = 3.7\sqrt{2}[R + 6.88 \exp(0.0072R - 0.000016R^2)]^{-1/2},$$

where

$$z = \left(\frac{\pi(\chi - 1)}{\chi} t_w \right)^{1/2},$$

$$Re_{0\text{eff}} = 10^{-m} Re_0, \quad m = 1.8(1 - h)^3,$$

$$R = Re_0 \left(\frac{3}{4} t_w + \frac{1}{4} \right)^{-0.67},$$

and h is a parameter of thinness of the apparatus, equal to the ratio of its height and length. The prescription of a surface is realized just in a same way, as it was done in Sec. 8.1.

The methodics proposed has well recommended itself in respect to the computation of hypersonic flows of the convex, not extremely thin and nonplanar pointed bodies. The computations are completely reflecting the qualitative behavior of C_x in dependence on the degree of rarefaction of the medium within the whole range of the angles of attack and provides the quantitative correspondence with the accuracy about 5% (Fig. 8.5).

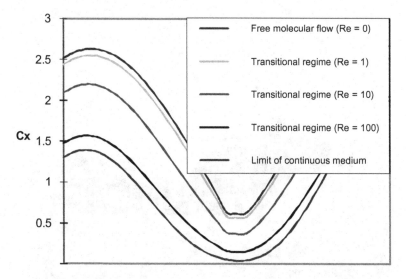

Fig. 8.5   Drag of ASS of the type "Shuttle" at the various stages of a trajectory.

8.4   The Results of the Flow About a Hypersonic Flying Apparatus "Clipper" (see Voronich, Zey Yar[225])

At the present time in Russia exists the promising project, of which the economical quality is in several times excelling the modern systems of the type of "Shuttle", and it is the returnable flying apparatus "Clipper", see Fig. 8.6. The economical advantage is reached through the technological simplicity and the comparatively small aerodynamical quality, which would be capable to draw the apparatus to the place of its landing. The landing itself is realized with the help of a parachute. The prehistory of the development of "Clipper" is connected with the space flying apparatus "Bohr" which were used during the development of "Bouran" system, for the determination of the main aerodynamical characteristics and the thermal loads. The first, nonpiloted start of "Clipper" is planned to be realized not later than in the year 2011, and the first piloted one — in 2012. The apparatus is projected with the following characteristics:

- The starting mass 13–14.5 ton.
- The number of the crew's members up to five persons.
- The volume of a cabin $20\,m^3$.
- The mass of load $500\,kg$.
- The speed up to $M = 20$.
- The range of the angles of attack from 10 to 40°.

Fig. 8.6   Model of ASS "Cliper" at the Aviasalon MAX-2007.

As for its construction, "Clipper" consists of the returnable apparatus (RA) and of the orbital section. The aerodynamical properties should permit to realize the gliding way of lowering in the upper layers of the atmosphere, the smooth maneuvre towards the point of landing, and all that will lower the thermal loads and will permit to use the multitemporal thermal protection. The RA is aerodynamically unstable, and for this reason with the aim of keeping the necessary orientation foreseen there are the aerodynamical shields. The project has also the versions of RA with the high aerodynamical quality at the subsonic regimes (up to the values of 4–5), and the presence of such a property permits for this apparatus to realize the landing at the air field just like for the usual airplane.

The development of HFA is complicated by the fact that there exists the whole series of the reproduction of natural conditions of flight in the wind tunnels. In particular, it is practically impossible to reproduce the thermal regime of the flow about the apparatus: the heating of a model within the tunnel leads to a high value of the temperature ratio, while in the natural conditions the temperature of a surface proves to be considerably lower than the total temperature of flow. The modeling of the high-speed flows suggests the observance of some other criteria of similitude, first of all concerning the numbers of Mach and Reynolds, as well as the provision of a low degree of turbulence and of uniformity of a flow in the operating part of a system. The serious influence of the accuracy of experiment is rendered also by the way of the model's fixation. The simultaneous solution of all these problems within the frame of a single experimental plant seems to be impossible. Therefore, for the study of high-speed flows are used the wind tunnels of various constructions and possessing the various principles of action.

The factors listed above stipulate the necessity of using the computational information at the stage of projection of the high-speed flying apparatus. The detailed parametrical studies at that stage might be carried out with the help of the complexes of the numerical modeling. The role of these might be played by the sets of program realizing the various models of the motion of medium. Among the complexes of that type might be considered the engineering computational programs, which use the hypothesis of a locality for high-speed flows. As the examples one could mention the complexes "MARK-IV" and "Height", which were developed in TSAGI (Central Aerodynamical Institute, named after Prof. N.E. Zhukovskii). With the help of these complexes one is able to obtain very quickly the large volume of information on the aerodynamical characteristics of the FA. The simplifying assumptions, on which are based such complexes, lead to the limitations on the accuracy of results and on the area of their application, and, therefore, for the improvement of the accuracy of results it should be necessary to turn to the more complete models, that is, to the equations by Euler, Navier–Stockes, and Boltzmann. The outstanding achievement of the last years is presented by the ITPM (Institute of Theoretical and Applied Mechanics) under the leadership of Ivanov[224] of the unique computational complex "SMILE". This complex permits to calculate the aerodynamical characteristics of the bodies of arbitrary configuration at the transitional regime, on the basis of the solution of Boltzmann equation by Monte Carlo method, without any simplifying assumptions.

The aim of the present study consists in determination of the main aerodynamical characteristics of one version of the complex "Clipper" by its hypersonic flight in the rarefied atmosphere, with the help of Monte Carlo method and the method, based on the hypothesis of locality.

Let us consider the application of a method to the solution of a problem of the determination of the aerodynamical characteristics of the body of a complicated form (the promising version of the creation of a returnable flying apparatus). The geometry of "Clipper" is presented by a set of triangles, while the triangulation of the body's surface and its reorganization into the format Stereo Lithography (STL) was carried out with the help of a complex of automatical protection SolidWorks, starting from the initial model imported from the IGS format. The format STL proves to be sufficiently simple and supports the property of a solid body.

The geometry of the considered version of RFA "Clipper" was represented by 3218 triangles, and the way of presentation is given in Fig. 8.7.

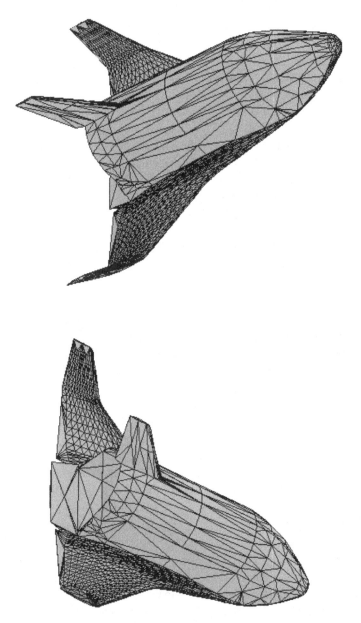

Fig. 8.7   Geometrical presentation of a version of setup of ASS "Cliper".

In the free-molecular case, the computational area presents in itself the rectangle with sizes $(x \times y \times z)$ $4l \times 2l \times 2l$, where l is the maximal external size of a body. The computations were carried out in the range of angles of attack from $-90°$ to $90°$, with a step of $5°$. The angle of attack was changed by way of rotation of a body about the axis z, around the centre of mass. The parameters of the problem are the following: velocity ratio $s = V_\infty/\sqrt{2RT_\infty} = 15$, adiabata index $\gamma = 5/3$, the values of temperature ratio $t_w = T_w/T_0$ were equal to 0.0004, 0.4. It was assumed that the reflection from the surface is a diffuse one. The computations were executed with the use of 10^6 particles, and were carried out as taking into account the multiple reflection of particles, then neglecting this factor.

Presented in Fig. 8.8 are the dependences of the coefficients of drag force C_x on the value of the angle of attack α, by the various values of the temperature ratio t_w. As it is clearly seen from these graphs, by the increase of the temperature ratio the body's drag is growing, and this fact is explained by the growth of contribution

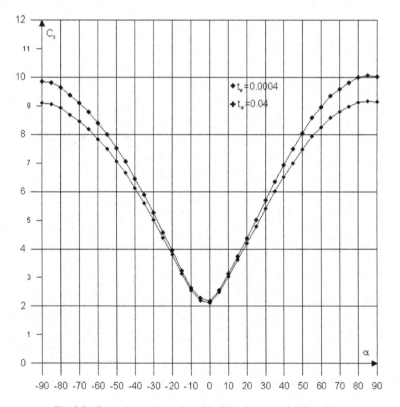

Fig. 8.8 Dependence $C_x(\alpha)$ for ASS "Cliper" at $t_w = 0.0004$, 0.04.

to the momentum of that from the reflected particles. Moreover, by the growth of a temperature ratio increased is the role of the body's form: for the "cold" body the graph of $C_x(\alpha)$ appears as almost symmetrical one, while for the heated body the nonsymmetry is more perceptible.

Presented in Fig. 8.9 are the dependence of the coefficient of lift force C_y on the value of the angle of attack α, by the various values of the temperature ratio t_w. As it follows from these results, with the increase of the temperature ratio is essentially increased the value of lift, and this is also explained by the growth of contribution to the momentum of that from the reflected particles. By $t_w = 0.04$ and $\alpha = 40°$ the coefficient C_y reaches the value of ~ 0.5, and this proves to be perceptible contribution to the balance of forces, acting on the body under consideration. Taking into account the multiple reflections leads to a notable increase of the absolute value of C_y by the negative angles of attack, see Fig. 8.9.

Presented in Fig. 8.10 are the dependence of the coefficient of moment of tangage m_z on the value of the angle of attack α, by the various values of the

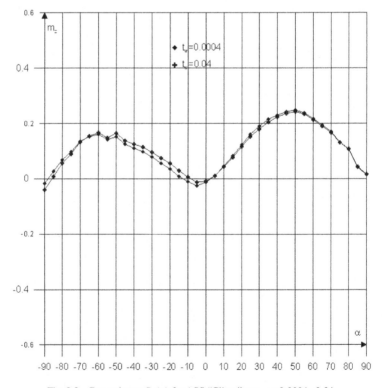

Fig. 8.9   Dependence $C_y(\alpha)$ for ASS "Cliper" at $t_w = 0.0004$, 0.04.

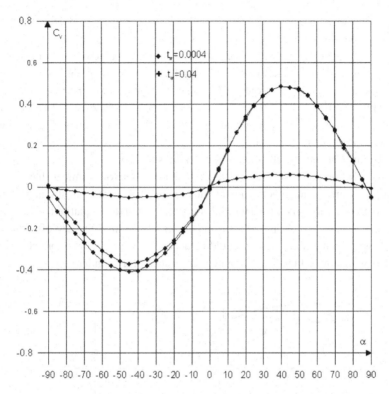

Fig. 8.10   Dependence $m_z(\alpha)$ for ASS "Cliper" at $t_w = 0.0004, 0.04$.

temperature ratio t_w. As it follows from these results, the coefficient of moment of tangage m_z is less sensitive to the variations of the temperature ratio, but its value means that this factor should be necessarily taken into account by the analysis of the variations of the body's orientation, when this body is subject to the action of a flow of strongly rarefied gas. Also, it is important that m_z is almost always positive, in distinction of the cases of circular and elliptical cones. The negative values of m_z by $t_w = 0.04$ are appearing in the range $-7° < \alpha < 3°$. By $\alpha < -7°$ the apparatus is unstable in respect to a tangage, and with increase of the temperature ratio the area of stability in respect of a tangage is narrowing, if $\alpha < 0°$.

One is ready to say something concerning the accuracy of the relations entering the method of locality. It is clear that these relations would occur to be applicable with the minimal error in the case of bodies with the form close to that of a sphere, and are not applicable in the case of very thin bodies, when the condition $M_\infty \sin\theta \gg 1$ is not observed. The methods considered are not taking into account the influence of the interaction of a boundary layer with the inviscid hypersonic flow at the large numbers of Re_∞. The computed and experimental values of C_x for a cone in the transitional regime are in a satisfactory agreement between themselves, while the data on C_y are in much worse agreement. It is necessary to note that the methodics proposed reflects qualitatively correctly the nonmonotony of a dependence of C_y for a cone on Re_∞. The computed and experimental results on C_x by $\alpha = 10°$ and $15°$ for a plate are in the good agreement, but the data on C_x by $\alpha = 5°$ and on C_y stay in bad agreement. This phenomenon is a sequence of the influence of interaction of a boundary layer with inviscid flow, which was not taken into account in the method of localities (Fig. 8.11).

Thus, the method of locality for the computation of aerodynamical characteristics of the bodies in the hypersonic flow of a rarefied gas, in the transitional regime, leads to good results on C_x for a wide class of bodies and to the qualitatively correct results on C_y. By the small angles of attack ($\alpha < 5°$) the accuracy of results is getting worse, and in this case it would be necessary to turn to the more complete models, which are taking into account the presence of a boundary layer.

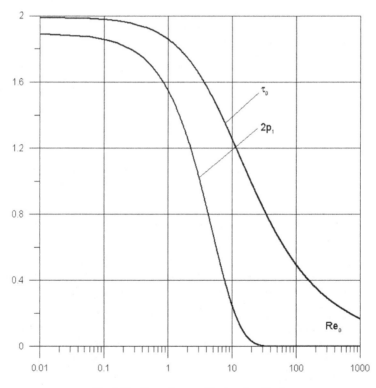

Fig. 8.11   Dependence $\tau_0(Re_0)$ and $p_1(Re_0)$.

The computations were carried out in the range of the angles of attack α and angles of slide β from $-90°$ up to $90°$ with a step of $3°$. The variation of the angle of attack was realized by means of rotation about the center of mass and around the z-axis, and the angle of slide was changed by means of rotation around the y-axis. The parameters of a problem were the following: specific heat ratio $\gamma = 1.4$, temperature ratio $t_w = T_w/T_0 = 0.001, 0.04, 0.1$, number by Reynolds $Re_0 = 0; 1; 10; 100; 1\,000; 10{,}000$.

Presented in Figs. 8.12(a)–8.12(c) are the dependence of the coefficient of drag force C_x on the value of angle of attack α, by the various value of Reynolds number and of the temperature ratio t_w. As it is evident from these results, with the increase of Reynolds number the coefficient of drag is decreasing (this tendency might be explained by the decrease of normal and tangential stresses, $p_1(Re_0)$ and $\tau_0(Re_0)$), while the general character of the dependence $C_x(\alpha)$ does not change. Let us note that the nosing part of an apparatus is blunted. The increase of a temperature ratio leads to the increase of body's drag, but by $Re_0 > 10^3$ the dependence $C_x(\alpha)$ ceases to be sensitive in respect to that parameter due to the decrease of $\tau_0(Re_0)$ and $p_1(Re_0)$. By the small values of temperature ratio t_w and by $Re_0 \ll 1$ the dependence $C_x(\alpha)$ proves to be almost symmetrical in respect to the ordinate axis.

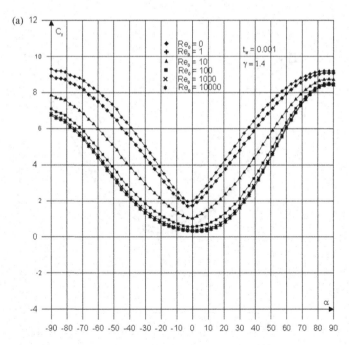

Fig. 8.12   Dependence $C_x(\alpha)$ for ASS "Cliper" at (a) $t_w = 0.001$; (b) $t_w = 0.04$; and (c) $t_w = 0.1$.

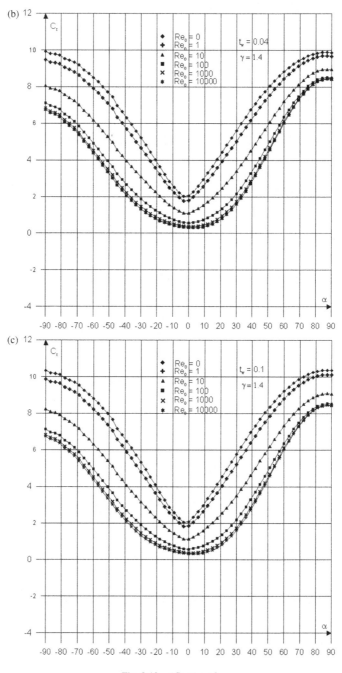

Fig. 8.12   (*Continued*)

Presented in Figs. 8.13(a)–8.13(c) are the dependences of the coefficient of lift force C_y on the value of angle of attack α, by the various values of Reynolds number and of the temperature ratio t_w. As it is evident from these results, the increase of Reynolds number leads to the increase, in the modular measurement, of the coefficient of lift, and the pique values of C_y by $Re_0 = 10^4$ and $t_w = 0.04$ are surpassing the pique values of C_y by $Re_0 = 0$. This is also explained by the character of behavior of the functions $\tau_0(Re_0)$ and $p_1(Re_0)$ at $Re_0 \to \infty$. The growth of a temperature ratio for the regimes, close to the free-molecular one, leads to a considerable growth of the absolute value of C_y and to the loss of symmetry of $C_y(\alpha)$ in respect to the coordinate origin, and these tendencies were noted to by analysis of the results obtained by Monte Carlo method. By $Re_0 > 10^3$, the dependence $C_y(\alpha)$ ceased to be sensitive in respect to the value of t_w. By $Re_0 \to \infty$ the dependence $C_y(\alpha)$ proves to be nonsymmetrical, and the modular value of C_y for the positive values of the angle of attack is essentially larger than the similar value of C_y for negative values of that angle. It is to be noted that the balancing value of the angle of attack by $Re_0 \to \infty$ is equal to $\alpha_0 \approx 3°$, and therewith one has $dC_y/d\alpha = 2.5 \times 10^{-2}$ 1/grad.

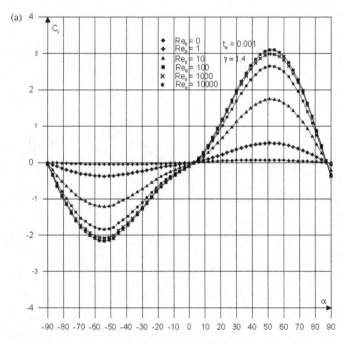

Fig. 8.13   Dependence $C_y(\alpha)$ for ASS "Cliper" at (a) $t_w = 0.001$; (b) $t_w = 0.04$; and (c) $t_w = 0.1$.

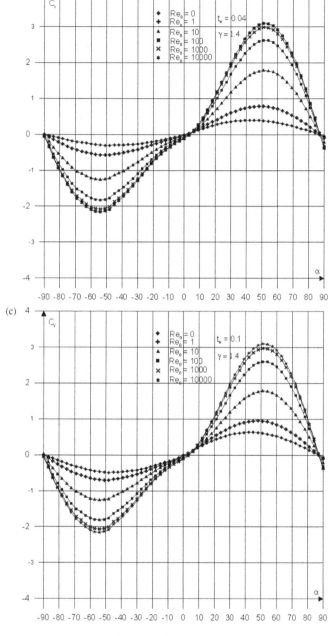

Fig. 8.13   (*Continued*)

Presented in Figs. 8.14(a)–8.14(c) are the dependences of the coefficient of moment of tangage m_z (in respect to the center of mass of the uniform body) on the value of the angle of attack α, by the various values of the Reynolds number and of the temperature ratio t_w. It is seen from these graphs that the coefficient of the moment of tangage m_z is poorly sensitive to the value of a temperature ratio, but rather sensitive to the value of Reynolds number. By the increase of Reynolds number occurs the change of sign of m_z for positive angles of attack, and the boundary value is that of $Re_0 \sim 10$. At $Re_0 \to \infty$ the pique value of $m_z = -0.14$ by the positive angles of attack is reached by $\alpha \approx 40°$. Of the practical interest is the change of sign of m_z within the range $-10° < \alpha < 3°$, by $Re_0 \to \infty$ and $t_w = 0.04$. Let us note that at $Re_0 \to \infty$ one has $m_z(\alpha) < 0$ by $\alpha > \alpha_0$ and $m_z(\alpha) > 0$ by $\alpha < \alpha_0$, which means that the apparatus is completely unstable in respect to a tangage.

Presented in Figs. 8.15(a)–8.15(c) are the dependences of the coefficient of drag force C_x on the values of the angle of slide β, by the various values of Reynolds number and of the temperature ratio t_w. The information of that type is also necessary for the obtainment of complete notion of the forces acting on the apparatus. In this case the regularities prove to be similar to the dependence $C_x(\alpha)$,

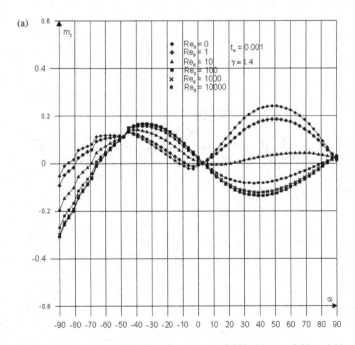

Fig. 8.14   Dependence $m_z(\alpha)$ for ASS "Cliper" at (a) $t_w = 0.001$; (b) $t_w = 0.04$; and (c) $t_w = 0.1$.

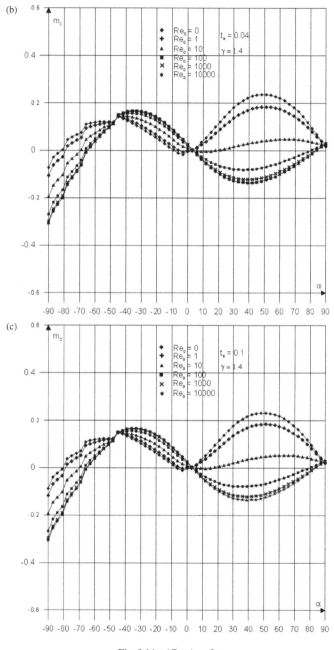

Fig. 8.14   (*Continued*)

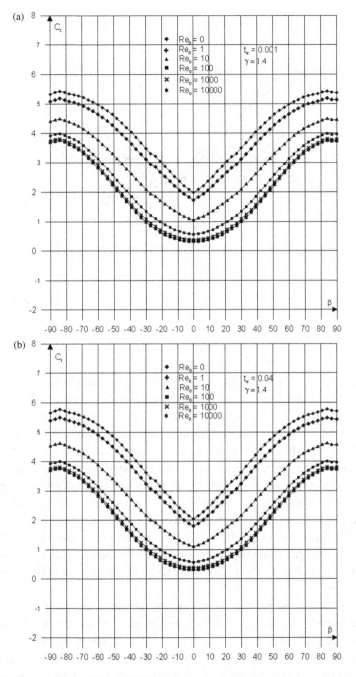

Fig. 8.15   Dependence $C_x(\beta)$ for ASS "Cliper" at (a) $t_w = 0.001$; (b) $t_w = 0.04$; and (c) $t_w = 0.1$.

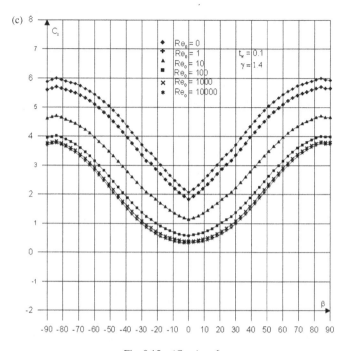

Fig. 8.15   (*Continued*)

with such a difference that $C_x(\beta)$ by $\beta = \pm 90°$ is less than $C_x(\alpha)$ by $\alpha = \pm 90°$. Moreover, the dependence $C_x(\beta)$ is strictly symmetrical in respect of the axis of ordinates due to the symmetry of the apparatus in respect to the plane.

Presented in Figs. 8.16(a)–8.16(c) are dependencies of the coefficient of side C_z force on the values of the angle of slide β, by the various values of Reynolds number and of the temperature ratio t_w. Due to the symmetry of the apparatus the dependence $C_z(\beta)$ possesses symmetry in respect to the axis of ordinates. By the increase of Reynolds number takes place an essential increase of the module of a side force's coefficient, and the pique values of C_z by $Re_0 = 10^4$ and $t_w = 0.04$ surpass the pique values of C_z by $Re_0 = 0$ in one order of magnitude. The growth of a temperature ratio leads to the considerable increase of the absolute value of C_z by $Re_0 = 0$. By $Re_0 > 10^3$ the dependence $C_z(\beta)$ ceases to be sensible to variations of t_w.

Presented in Figs. 8.17(a)–8.17(c) are the dependences of the coefficient of moment of rush m_y (in respect to the center of mass of a uniform body) on the values of the angle of slide β, by the various values of the Reynolds number and of the temperature ratio t_w. It is clearly seen from these graphs that the coefficient of moment of rush m_y is poorly sensitive to the value of Reynolds number. By the

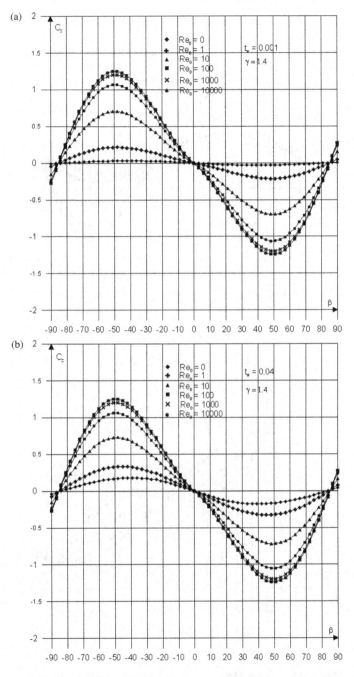

Fig. 8.16   Dependence $C_z(\beta)$ for ASS "Cliper" at (a) $t_w = 0.001$; (b) $t_w = 0.04$; and (c) $t_w = 0.1$.

(c)

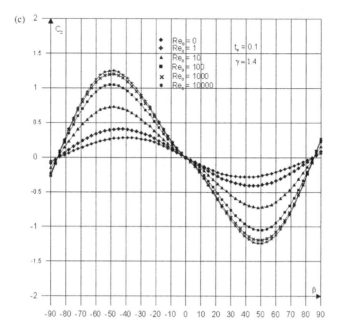

Fig. 8.16   (*Continued*)

(a)

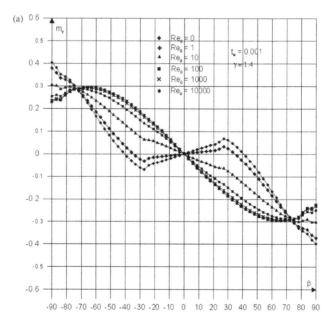

Fig. 8.17   Dependence $m_y(\beta)$ for ASS "Cliper" at (a) $t_w = 0.001$; (b) $t_w = 0.04$; and (c) $t_w = 0.1$.

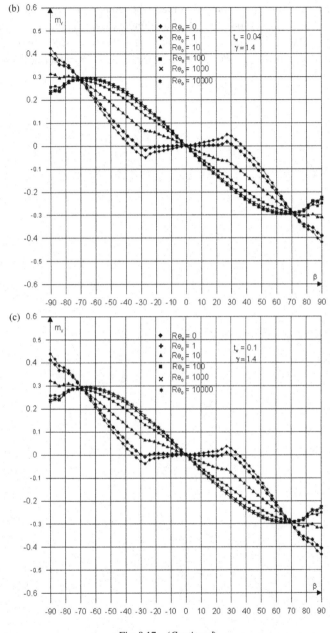

Fig. 8.17 (*Continued*)

increase of the Reynolds number takes place the change of sign of m_y in vicinity of $\beta = 0°$, and by $t_w = 0.04$ the role of a boundary value is played by $Re_0 \sim 1$. By $Re_0 \to \infty$ the pique value $m_y = 0.3$ is reached by $\beta = -65°$. Let us note that by $\beta = -65°$ one has $dm_y/d\beta < 0$ in vicinity of $\beta = 0°$, which means that the apparatus is unstable in respect of rush.

Chapter 9

The Flow About Blunted Bodies with the Addition of Heat (see Vorovich, Moiseev)

9.1 The Main Features of a Method

Revealed in the preceding chapters were possibilities of computations of the flows of rarefied gases within the wide range of the degrees of rarefaction, on the basis of Boltzmann equation,

$$\frac{\partial f}{\partial t} + \vec{\xi} \nabla f = I(f), \tag{9.1.1}$$

using the scheme "relaxation–transfer", which is splitting the process of solution at the step Δt into two independent stages:

$$\frac{\partial f}{\partial t} + \vec{\xi} \nabla f = 0, \quad \frac{\partial f}{\partial t} = I(f). \tag{9.1.2}$$

This splitting corresponds to the collisionless transfer and to the spatially uniform relaxation, correspondingly (here symbol $I(f)$ means the integral of collisions). Using the moment approximations of the distribution function:

$$f = F(\vec{\xi}, M^0, M^1, M^2, \ldots, M^K), \tag{9.1.3}$$

the stages of transfer and of relaxation might be realized at the level of moments, also:

$$\frac{\partial M^k}{\partial t} = \alpha M^k, \tag{9.1.4}$$

where $\alpha = 0$ for the moments $M^k = n, \vec{u}, T$ (conservation of the mass, momentum, and energy at the stage of relaxation), and $\alpha = -1/\tau_r$ for the stress tensor, $\alpha = -2/3\tau_r$ for the vector of heat flow, τ_r — time of relaxation. The stage of transfer might be realized either at the level of distribution function, with the using of the modeling particles or without it, or at the level of moments, using the various approaches (obtained are the sets of quasi-hydrodynamical equations). Such a method permits to realize the modeling of the flows of a continuous medium, where the external form of a distribution function is known — it might be the distribution function by Euler, by Navier–Stokes, or by Burnett.

In the work by Pullin[81] is carried out the modeling of flows of the inviscid, compressible, ideal gas, with the help of a locally equilibrium distribution function f_0:

$$f_0 = \left(\frac{m}{2\pi kT}\right)^{3/2} \exp\left(-\frac{m(\vec{\xi} - \vec{u})^2}{2kT}\right),$$

either at the level of particles, or at the level of macroscopic quantities. Considered is the one-dimensional problem of the formation of a shock in front of a piston, which is moving with a supersonic speed relatively to a gas in rest. Presented below is the description of an algorithm, realizable at the level of particles, and the discussion of the problem of choice of steps in time and in space. In the present chapter computed are the flows of inviscid, compressible, diatomic ideal gas, described by the locally equilibrium distribution function f_0. Used for computations is the scheme "relaxation–transfer", described above. Computations are carried out at the level of particles. In this case, the algorithm might be presented in the following way. The ensemble of particles with velocities $\vec{\xi}_i$ and with internal energies ε_i is in motion during the temporal intervals Δt, and after each of these intervals are computed the macroparameters within the spatial cells, $\Omega = \{(k, l), k = 1, \ldots, K, l = 1, \ldots, L\}$, which have a size $\Delta x \times \Delta y$. In these processes, during the non-stop flights are taken into account the boundary conditions (the specular reflection of particles from the surface, which would provide the condition of slip, the condition of local uniformity or of the absence of perturbations at the free boundary). The choice of a temporal step Δt is realized, proceeding from the smallness of variations of the distribution function in the present point of space:

$$\frac{\partial f}{\partial t} \Delta t \ll f. \tag{9.1.5}$$

More concretely, in the general three-dimensional case one has

$$\Delta t = s \min(\Delta x, \Delta y, \Delta z) / \max(U + a_x),$$

where $s < 1$, a_x — the speed of sound. The value of spatial step is chosen proceeding from the gradient of macroparameters. Thus, the choice of this step is corresponding to the continuous medium. For the computations of macroparameters within the cell (k, l) are used the formulae:

$$\vec{U}_{kl} = \frac{1}{N_{kl}} \sum_{i=1}^{N_{kl}} \vec{\xi}_i,$$ (9.1.6)

$$T_{kl} = \frac{2}{(3 + \nu)k_B} \frac{1}{N_{kl}}$$

$$\times \sum_{i=1}^{N_{kl}} \left\{ \frac{m}{2}(\xi_{ix} - U_{klx})^2 + \frac{m}{2}(\xi_{iy} - U_{kly})^2 + \frac{m}{2}\xi_{jz}^2 + \nu \frac{k_B T_i''}{2} \right\}.$$

9.2 Description of the Algorithm

Nonstop flight in the two-dimensional case is realized according to the formulae

$$x_{i2} = x_{i1} + \xi_{ix}'' \cdot \Delta t,$$

$$y_{i2} = y_{i1} + \xi_{iy}'' \cdot \Delta t.$$

Here the progressive degree of freedom, the macroscoping motion in accordance with which is absent, might be included into expression of the internal energy by way of adding the item $\frac{k_B T_i''}{2}$,[81] the rotation degrees of freedom — by way of addition the item $\nu \cdot \frac{k_B T_i''}{2}$,[112] where T'' is the temperature in the initial cell for the particle considered, N_{kl} — the number of particles within the cell (k, l) at the present temporal step t, ν — the number of internal degrees of freedom in a molecule. After the computation of macroparameters for each particle is gambled the velocity, corresponding to the distribution function f_0 with new values of macroparameters.

One of the peculiar features of the algorithm consists in the fact, that the coordinates of particles are not memorized, but are uniformly distributed over the volume of a cell before each stage of transfer, and thus are economized the computational resources (the operative memory). Further on, it will be shown that this feature does not influence the order of approximation of the method.

At the stage of getting of the ensemble of velocities on the basis of a distribution function f_0 arises the necessity to obtain the ensemble N_{kl} of particles within the cell (k, l) with velocities ξ_i, $i = 1, \ldots, N_{kl}$, corresponding to a function f_0 with new macroparameters. This stage is carried out in accordance to the theory of the mathematical statistics, by way of solution of the problem of obtainment of a finite set of the accidental quantities ξ_i using the distribution function f_0. The most widespread way of solution of this problem is the following. One considers the accidental quantities η and θ, which possess the densities of probability

$$p_\eta(x) = \left(\frac{\beta}{\pi}\right)^{1/2} \exp(-\beta x^2), \quad p_\theta(y) = \left(\frac{\beta}{\pi}\right)^{1/2} \exp(-\beta y^2),$$

correspondingly. If one considers the joint distribution of these quantities

$$p_Q(x, y) = p_\eta(x) p_\theta(y) = \left(\frac{\beta}{\pi}\right) \exp(-\beta(x^2 + y^2)),$$

and rewrites this expression in the polar coordinates

$$x = r \cdot \cos \phi,$$

$$y = r \cdot \sin \phi,$$

then it would be easy to obtain the following expression:

$$p_Q(r, \theta) = r \left(\frac{\beta}{\pi}\right) \exp(-\beta r^2).$$

It is not difficult to note that the function $p_Q(r, \varphi)$ is just a product of two standardized functions:

$$p_Q(r, \phi) = \frac{1}{2\pi} \cdot 2r\beta e^{-\beta r^2} = p_\phi \cdot p_r,$$

where

$$p_\phi = \frac{1}{2\pi}, \quad p_r = 2r\beta e^{-\beta r^2}.$$

Thus, in accordance to the theory of mathematical statistics one might write down the relations between the unknown accidental quantities r_Q and φ_Q and the independent accidental quantities α_1 and α_2, which are uniformly distributed within the interval $(0, 1)$:

$$\int_0^{r_Q} 2r\beta \exp(-\beta r^2)dr = 1 - \alpha_1, \quad \int_0^{\varphi_Q} \frac{1}{2\pi}d\varphi = \alpha_2.$$

(In the present case in the right-hand side of the first relation is included for the convenience the accidental quantity $(1 - \alpha_1)$, but not α_1.)

Ultimately, one has

$$r_Q = \frac{1}{\sqrt{\beta}}\sqrt{-\ln \alpha_1}, \quad \phi_Q = 2\pi\alpha_2.$$

Returning to the initial Cartesian coordinates, one obtains the expressions for η and θ:

$$\eta = \frac{1}{\sqrt{\beta}} \cdot \sqrt{-\ln \alpha_1} \cdot \cos(2\pi\alpha_2), \quad \theta = \frac{1}{\sqrt{\beta}} \cdot \sqrt{-\ln \alpha_1} \cdot \sin(2\pi\alpha_2).$$

Using the similar algorithm, it would not be difficult to obtain the formulae for the three components of velocity, distributed in accordance to the function f_0:

$$\xi'_{i1} = \sqrt{\frac{2kT}{m}} \cdot \sqrt{-\ln \alpha_{i1}} \cdot \cos(2\pi\alpha_{i2}),$$

$$\xi'_{i2} = \sqrt{\frac{2kT}{m}} \cdot \sqrt{-\ln \alpha_{i1}} \cdot \sin(2\pi\alpha_{i2}),$$

$$\xi'_{i3} = \sqrt{\frac{2kT}{m}} \cdot \sqrt{-\ln \alpha_{i3}} \cdot \cos(2\pi\alpha_{i4}),$$

$$\vec{\xi}_i = \vec{\xi}'_i + \vec{U}, \quad i = 1, \ldots, N.$$

Here α_{ik} are the independent accidental numbers uniformly distributed at the interval $(0, 1)$.

One of the peculiar features of the method proposed appears to be the fact, that due to the limitation on the number of molecules (in the computations reported were used from 50 and up to 6 000 molecules in the cell, in the unperturbed flow) the condition of conservativeness is not observed. Thus, at the stage considered it is necessary to carry out the transformation of the velocities gambled. Let us calculate the following quantities:

$$\frac{1}{N} \sum_{i=1}^{N} \vec{\xi}'_i = \delta\vec{U}, \quad \frac{1}{N} \sum_{i=1}^{N} (\vec{\xi}'_i)^2 = 3RT'.$$

The velocities of molecules are calculated according to a new formula:

$$\vec{\xi}''_i = (T/T')^{1/2}(\vec{\xi}'_i - \delta\vec{U}) + \vec{U}.$$

Now, in this case the condition of conservativeness is observed:

$$\frac{1}{N} \sum_{i=1}^{N} \vec{\xi}''_i = \vec{U}, \quad \frac{1}{N} \sum_{i=1}^{N} (\vec{\xi}''_i - \vec{U})^2 = 3RT.$$

When carrying out the numerical realization, one might reduce the period of gambling of velocities by way of using the symmetrized algorithm. This algorithm presumes, that calculated is only one half of all the velocities, namely — calculated are only $\vec{\xi}'_{2i-1}$, that is, velocities of the particles with uneven indices. Velocities of the particles with even indices are taken as symmetrical in respect to calculated ones $\vec{\xi}'_{2i} = -\vec{\xi}'_{2i-1}$.

In such a case, when the cell contains uneven number of particles, one might use one of two methods: either to choose the number of particles within the cell to be even one, using for that the proper weight, or to set for the particle left the velocity being equal to zero, $\vec{\xi}'_N = 0$. In any of these cases will be observed the relation

$$\frac{1}{N}\sum_{i=1}^{N}\vec{\xi}'_i = 0.$$

Thus by the computation of $\vec{\xi}''$ it is taken that $\delta\vec{U} \equiv 0$.

9.3   The Approximational Properties

The estimation of approximational properties of the method of direct statistical modeling (DSM) for the inviscid flows was realized on the basis of expressions for flows of the form

$$F_{\varphi}^{\vec{n}} = \rho \int_{(\vec{\xi},\vec{n})>0} f_0\tilde{\varphi}(\vec{\xi},\vec{n})d\vec{\xi}, \qquad (9.3.1)$$

where $F_{\varphi}^{\vec{n}}$ is a flow of quantity $\tilde{\varphi}$ in the direction of normal $\vec{n}$, f_0 is a locally equilibrium function, $\tilde{\varphi} = \left\{1, \vec{\xi}, \frac{\xi^2}{2}\right\}$.

In the one-dimensional case the equation of balance for $\varphi(\varphi = m\tilde{\varphi})$, presented in the uniform system of cells will have the following form:

$$\frac{1}{\Delta V}\sum_j \varphi_j(t + \Delta t)$$

$$= \frac{1}{\Delta V}\sum_j \varphi_j(t) + \left[-F_{\phi}^{ln+} + F_{\phi}^{l+1n-} - F_{\phi}^{ln-} + F_{\phi}^{l-1n+}\right]\Delta t/\Delta x, \quad (9.3.2)$$

where $n+$ is a normal vector in the positive direction, $n-$ — a normal vector in the negative direction, l — a number of cells in the network $\Omega = \{\omega_l, l = 1, \dots, L\}$, which has the size Δx, $\omega_l = (x_l - \Delta x/2, x_l + \Delta x/2)$, ΔV — the volume of a cell. Expressions (9.3.1) and (9.3.2) were used in the paper by Pullin[81] for the computations connected with a problem of piston with the help of macroparameters, and in the present book — with the help of Monte Carlo method. Actually, these expressions correspond to the method of differences. The approximation of the stage of transfer was realized by the use of expressions (9.3.1) and (9.3.2) is $O(\Delta t, \Delta x)$.

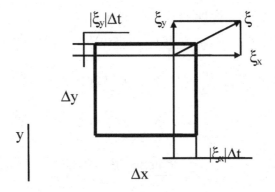

Fig. 9.1

In the one-dimensional case this result might be simply obtained from (9.3.2). The execution of similar computations for the two-dimensional case is more cumbersome not only due to the presence of additional dimension, but also because of the necessity of a special consideration of the cell's angles. As a whole, the schemes of the type of (9.3.2) might be used in the difference methods of gas dynamics. The difficulties in the use of these schemes are diminished by the complicated nets in the computational area, and in such conditions the realization of computations at the level of particles might be really justified.

The consideration of approximational properties might be carried out directly on the basis of algorithm. By the use of a uniform rectilinear net (Fig. 9.1) the problem looks as follows (for the simplicity and clearness is taken the two-dimensional case). For the particles of the main group $(u_x \pm a_s) \times (u_y \pm a_s)$, by the indicated earlier limitation on the value of step Δt, the exchange between the cells, for the particles belonging to the group $\vec{\xi}$, will be realized in the layers $|\xi_x|\Delta t < \Delta x$, $|\xi_y|\Delta t < \Delta y$.

In the method of DSM for a rarefied gas the choice of a temporal step is different: $\Delta t \approx \min(\Delta x)/U_{\max}$, where $\Delta x \approx \lambda/3$, λ — the local length of a mean free path, and such a consideration, as mentioned earlier, is inapplicable.

The equation of balance in the two-dimensional case might be written down ($N^t_{ij\xi}$ — the number of particles of $\vec{\xi}$ in the cell (i, j) : $\left(x_i \pm \frac{1}{2}\Delta x\right) \times \left(y_j \pm \frac{1}{2}\Delta y\right)$ at the moment of time t) taking into account the uniform distribution of particles over the volume of a cell, and for this reason the fractions represent the relative (in respect to $N^t_{ij\xi}$) numbers of particles, which have flown out of the cell and those, which have flown into it from the neighboring cells. This consideration is independent of the Mach number and of the external form of a distribution function (with the exception of the case of a decrease by the velocities of large modules).

Equation (9.3.3) depicts the balance of particles.

$$N_{ij\xi}^{t+\Delta t} = N_{ij\xi}^t - N_{ij\xi}^t \frac{|\xi_x|\Delta t \Delta y + |\xi_y|\Delta t \Delta x - |\xi_x||\xi_y|(\Delta t)^2}{\Delta x \Delta y}$$

$$+ N_{i-\omega x j\xi}^t \frac{|\xi_x|\Delta t \Delta y - |\xi_x||\xi_y|(\Delta t)^2}{\Delta x \Delta y}$$

$$+ N_{ij-\omega y\xi}^t \frac{|\xi_y|\Delta t \Delta x - |\xi_y||\xi_x|(\Delta t)^2}{\Delta x \Delta y}$$

$$+ N_{i-\omega x j-\omega y\xi}^t \frac{|\xi_x||\xi_y|(\Delta t)^2}{\Delta x \Delta y}, \tag{9.3.3}$$

where $\omega x = \text{sign}(\xi_x)$, $\omega y = \text{sign}(\xi_y)$.

Upon assumption that there exists sufficiently smooth function $\tilde{f}$:

$$\tilde{f}(t, x_i, y_j, \vec{\xi}) = N_{ij\xi}^t / \Delta x \Delta y \Delta \vec{\xi},$$

after making the expansions in series in vicinity of a cell (i, j), one might present Eq. (9.3.3) in the form

$$\frac{\partial \tilde{f}}{\partial t} + \vec{\xi} \frac{\partial \tilde{f}}{\partial \vec{x}} = O(\Delta t, \Delta x). \tag{9.3.4}$$

There exists, moreover, the statistical error $\sim 1/\sqrt{N_{ij}}$, where N_{ij} is the number of particles within the cell. The statistical nature of the method proposed demands the realization of rather considerable temporal expenditures, needed for the obtainment of the solution with high accuracy, but the main features of a solution might be obtained for two-dimensional, and, especially, for three-dimensional problems, in duration of a temporal interval, comparable to that for the widespread gasdynamical methods. The choice of the spatial step and of the number of modeling particles ought to be in concordance with the choice of a parameter s, too. When the conservativeness of the DSM method is guaranteed, then the stationary solution, obtained with the small ensemble of particles, by the temporal averaging is made more precise in all the details, just as it would be obtained by the averaging over the ensemble. As a rather important problem appears that of the improvement of approximation, and in a close connection with that are the questions about the construction of nets and on the correction of flows serving to such an approximation.

9.4 The Algorithm and the Nets

Used by the computations of two-dimensional flows were the rectilinear nets. The sizes of nets for the two-dimensional problems were 90 × 60, and for the axisymmetrical ones — 90 × 35. At the upper boundary was set the condition of the unperturbed flow. The same condition was set at the frontal boundary, too. At the backside boundary was set the condition of the local equilibrium state of the flow. By the consideration of two-dimensional problems it was assumed that the flow possesses the unit width, and thus the volume of a cell proved to be numerically equal to its area in the (x, y) plane. By the consideration of axisymmetrical flows the cell's volume ought to be calculated as the volume of a sector of the corresponding ring. In this case volume of the cells, located at the larger distances from the axis, would be larger than the volume of a cell of the same area in the plane, but located closer to the axis. This means, that if the density of gas is one and the same within all the cells, then the number of molecules far from the axis is sufficiently large by the prescribed number of particles near the axis, which should be necessary for the obtainment of a satisfactory statistics. Thus the time, necessary for the computation, proves to be increased. For a solution of this problem introduced is the weighting factor. This factor leads to the obtainment of a result in rather crude approximation, but to the considerable decrease of a number of particles within the area, and, correspondingly, of the duration of computation. In view of the fact that the relaxational problem is not solved with the help of this method, the negative influence of the weighting factor seems to be not seriously essential one.[10] In the axisymmetrical problems considered here, by the obtainment of a crude approximation the number of molecules within the cell of the first layer in the unperturbed flow was set to be equal to 1000, while the number of molecules within the area was limited by a figure of 2000.

Moreover, by the consideration both of the two-dimensional and of the axisymmetrical flows the ultimate results were obtained, using the method of averaging. The essence of this method consists in that the results, obtained at the last several steps, are averaged. This permitted to get the acceptable accuracy even for the areas with small gradients, though such areas are poorly solved by the present method. The block-scheme of a program, which was used for the obtainment of results, is presented in Fig. 9.1.

By the calculation of macroparameters were used the nondimensional variables $\tilde{T} = \frac{T}{T_\infty}$ and $\tilde{U} = \frac{U}{\sqrt{RT_\infty}}$, where $R = \frac{k_B}{m}$, so that the equations for calculation of the temperature and of the molecular velocities acquire the form:

$$\tilde{T}_{kl} = \frac{1}{(3 + v) \cdot N_{kl}}$$

$$\times \sum_{i=1}^{N_{kl}} \left\{ (\tilde{\xi}_{ix} - \tilde{U}_{klx})^2 + (\tilde{\xi}_{iy} - \tilde{U}_{kly})^2 + \xi_{jz}^2 + v \cdot \tilde{T}_i'' \right\}, \quad (9.4.1)$$

$$\tilde{\xi}_{ij}' = \sqrt{2\tilde{T}} \cdot \sqrt{-\ln \alpha_{j1}} \cdot \cos(2\pi \alpha_{j2}). \quad (9.4.2)$$

In this case the velocity of the unperturbed oncoming flow is equal to $\tilde{U}_\infty = M_\infty \cdot \sqrt{\frac{5+\nu}{3+\nu}}$, where M_∞ is the Mach number.

Since in the present method are not considered the physical collisions between the molecules, then appears to be senseless the real physical size of a cell in its relation to the length of path, because the notion of the physical length of a free path of molecule is now absent. The limitations set on the relation of the cell's size and the interval of time Δt, were discussed earlier.

9.4.1   *Axisymmetrical case*

In the present analysis the computation of the axisymmetrical flows is reduced to that of the two-dimensional flows. The scheme of algorithm as a whole is kept unchanged, but there are some distinctions in the realization of the stage of transfer. Considered is the cartesian coordinate system, one of the axes of which coincides with the axis of symmetry of a flow (in our case it is x-axis). Prior to the stage of transfer all the particles within the area are uniformly distributed over the cell's volumes in the plane (x, y). After that, in accordance to the well-known formulae are gambled the velocities and are realized the nonstop flights of the particles in a three-dimensional space, and, finally, the new coordinates of particles are calculated:

$$x_{j2} = x_{j1} + \xi''_{jx} \cdot \Delta t,$$

$$y_{j2} = y_{j1} + \xi''_{jy} \cdot \Delta t, \qquad (9.4.3)$$

$$z_{j2} = z_{j1} + \xi''_{jx} \cdot \Delta t.$$

During the computation of macroparameters in the cells only two coordinates of the particles are taken into account, namely, (x_{j2}, y_{jN}), where $y_{jN} = \sqrt{y_{j2}^2 + z_{j2}^2}$. In the present case y_{jN} plays the role of radius, that is, of the distance from the final position of the particle to the axis of symmetry of a flow. And before the computation of macroscopical velocities within the cells the components of the velocities of particles, ξ''_{jy} and ξ''_{jz} should be transformed into the new components, ξ^*_{jy} and ξ^*_{jz} in such a way, that the quantity ξ^*_{jy} remains to be the radial component. The new values, ξ^*_{jy} and ξ^*_{jz} are determined as

$$\xi^*_{jy} = \left(\xi''_{jy}(y_{j1} + \xi''_{jy} \cdot \Delta t) + \xi''^2_{jz} \cdot \Delta t\right)/y_{jN}, \qquad (9.4.4)$$

$$\xi^*_{jz} = \left(\xi''_{jz}(y_{j1} + \xi''_{jy} \cdot \Delta t) - \xi''_{jy} \cdot \xi''_{jz} \cdot \Delta t\right)/y_{jN}. \qquad (9.4.5)$$

Let us divide the area into the totality of L cells along the axis x, so that $\Delta x_1 = \Delta x_2 = \cdots = \Delta x_L = \Delta x$. Consequently, the cell having number i is bounded by the coordinates $x_{i-1} = (i - 1) \cdot \Delta x$ and $x_i = i \cdot \Delta x$. In direction of the coordinate y the area is divided into M cells in such a way, that in any plane x each of the cells is bounded by two radii: the lower one, $r_{j-1}(x) = (j - 1) \cdot \Delta r$, and the upper one, $r_j(x) = r_{j-1}(x) + \Delta r$. In direction of the third coordinate, φ, the area is bounded by two planes, passing across the x-axis, and the angle between them is equal to $\Delta\varphi$. For the area of cross section of the arbitrary cell (i, j) by the plane $x = x^*$ one obtains the following expression:

$$S(x^*) = \frac{\Delta\varphi}{2} \cdot (2 \cdot r_{j-1}(x^*) + \Delta r) \cdot \Delta r, \qquad (9.4.6)$$

where $x^* \in [x_{i-1}, x_i]$.

The volume of a cell under consideration is

$$V_{ij} = \int_{x_{i-1}}^{x_i} S(x^*)dx = \frac{\Delta\varphi \cdot \Delta r}{2}((2j - 1) \cdot \Delta r) \cdot \Delta x. \qquad (9.4.7)$$

After the determination of x_k it is necessary to find the second coordinate of a particle, y_k. For the realization of this one should solve the problem on the uniform distribution of particles over the cross section of a cell by the plane $x = x_k$. This problem is solved similarly to the preceding one: the ratio of areas

$$S_k^{**}(\Delta r_k^{**}) = \frac{\Delta\varphi}{2}(2r_{j-1}(x_k) + \Delta r_k^{**}) \cdot \Delta r_k^{**} \qquad (9.4.8)$$

$$\text{and} \quad S_k = \frac{\Delta\varphi}{2}(2r_{j-1}(x_k) + \Delta r) \cdot \Delta r, \qquad (9.4.9)$$

gives the function of distribution of the accidental quantity η, determinating the particle's position along the coordinate y:

$$F_\eta(\Delta r_k^{**}) = \frac{S_k^{**}(\Delta r_k^{**})}{S_k} = \frac{2r_{j-1}(x_k) \cdot \Delta r_k^{**} + \Delta r_k^{**2}}{2r_{j-1}(x_k) \cdot \Delta r + \Delta r^2} \qquad (9.4.10)$$

(here, similarly to the preceding case, is used the notation $\Delta r_k^{**} = r_k^{**} - r_{j-1}(x_k)$, where r_k^{**} is a variable, belonging to the segment $[r_{j-1}(x_k), r_j(x_k)]$). Ultimately, one finds $y_k = \Delta r\sqrt{j^2 - (2j - 1)\lambda_y}$, where λ_y is the accidental quantity, uniformly distributed within the interval $(0,1)$. The particles are uniformly distributed along the coordinate x.

9.5   Direct Statistical Modeling of the Inviscid Flows About Blunted Bodies by the Presence of Energy Addition

As it was already noted in reference connected with the subject of the present paragraph,[126,127] by way of the proper location of the areas of heat addition near the external surface of the body in a flow one is able to diminish the wave drag and to obtain the controlling aerodynamical forces. Among the realistic approaches to the voluminous addition of a heat to the flow appears the absorption of a laser radiation and the use of the systems of electrical discharge.[127] Considered in the present work are the problems of the supersonic flows about two-dimensional blunted bodies (the plane butt, the cylindrical butt), in the case of a perfect gas and of the presence of the voluminously distributed source of a heat, located in front of a body.

The problems mentioned were solved with the help of a method of DSM for inviscid flows, which possesses the absolute stability, the practical monotony and the compactness of the computational pattern corresponding to the algorithm. This method, just as all the other methods of DSM, possesses the approximation of the first order, both in space and in time. The computation was carried out by the through method, up to the achievement of the temporal stability. The surfaces of shock waves were not singled out. There were not set any additional conditions at the boundary of a voluminous source. At the external boundaries were set the conditions of an unperturbed flow, at the axis or the plane of symmetry and at the backward boundary were set the conditions of zeroing of the normal derivatives of gasdynamical functions. At the body's surface was set the nonflow-through condition. The boundary conditions were realized at the stage of transfer in the algorithm of direct modeling. At the body and at the plane of symmetry the particles are reflected specularly, while at the other boundaries the necessary conditions were realized by a special way. Taking into account of the internal degrees of freedom was realized by way of a transfer of the averaged within the cell internal energy by each of the model particles. The internal energy was calculated with the assumption of equilibrium, and was assumed to be uniformly distributed over the particles within the cell considered.

Computed initially were the problems of a symmetrical flow about the plane and axisymmetrical butts, the flow being supersonic one, with $M_\infty = 3$, $\gamma = 1.4$. The problems of this type were repeatedly computed with the help of the various methods. The averaged correspondence of data was found within the limits of 3–5%. For the purpose of control was computed also the pressure at the stagnation point behind the straight shock, which might be obtained theoretically.[129]

The fields of flows obtained were used as an initial approximation for computation of the flows with heat addition. From the point of view of practical applications the main attention was attached to a cylindrical butt. The source of heat occupied several cells with a constant density of heat secretion. This source was located in the plane of symmetry or at the axis of symmetry, at some distance ahead of a body.

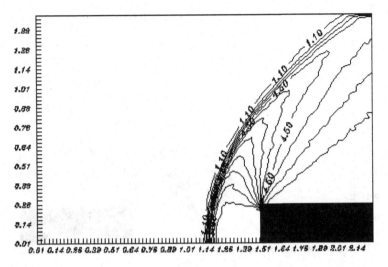

Fig. 9.2   Isolines of the pressure field ($M_\infty = 3$, $\gamma = 1.4$).

The intensity of that source was determined by a dimensionless parameter Q_0:

$$I = Q_0 \frac{n_\infty (3 + \nu)k_3 T_\infty}{2\Delta t},$$

and was kept constant in time. The source occupied 3–9 cells and had a fixed length. Moreover, investigated was also the influence of the cross-sectional size of a source by the fixed capacity.

Presented in Figs. 9.2–9.4 are some results of computations of the flow about plane butt. The main qualitative features of these results are the same, as in the case of a cylindrical butt. Further on, the computational results for the cylindrical butt will be discussed in more detail.

As it is known,[126] formed behind the heated spot is rather lengthy area of the strongly heated gas with a lowered density and pressure. At the periphery appears a pendant shock wave.

By the sufficiently intense addition of heat inside the heated spot appears a local subsonic zone. The losses of total pressure for those particles, which have passed through the spot, prove to be rather considerable and are increasing with the growth of the contribution of energy (see Fig. 9.5). By this process takes place the effect of a "satiation" of the flow by the heat: beginning with some threshold value of energetical contribution the total pressure ceases to decrease after reaching its minimal value, making up about 40% of the initial one, by the Mach number $M_\infty = 3$. The increased loss of a total pressure leads to the lowering of the wave drag of the bodies located downstream.[127] The presence of a heat addition exerts

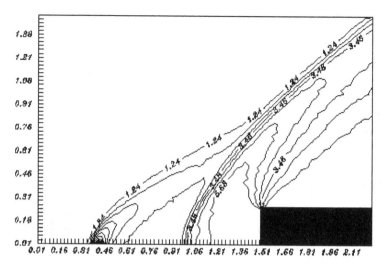

Fig. 9.3   Isolines of the pressure field for the flow with source ($M_\infty = 3$, $\gamma = 1.4$, $Q_0 = 4$).

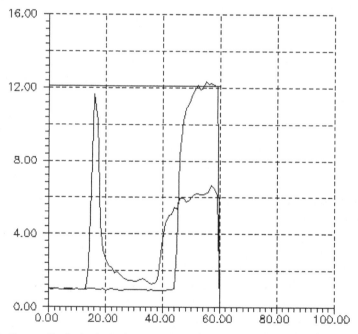

Fig. 9.4   Pressure distribution in the plane of symmetry with a heater and without it ($M_\infty = 3$, $\gamma = 1.4$, $Q_0 = 4$).

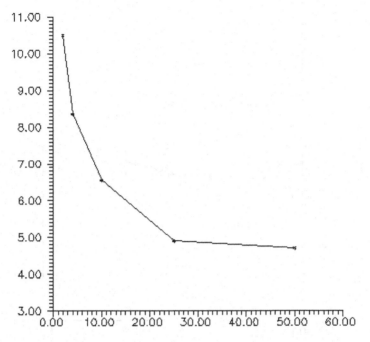

Fig. 9.5   Dependence of the stagnation pressure on the source's intensity.

a strong influence on the structure of a flow. The problem proves to be nonlinear one, the perturbations are strong, and as almost the only way of the problem's solution appears the numerical computation within the frames of gas dynamics. The flow, which has passed through the source, proves to be nonuniform one: those particles, which have passed through the spot, are heated more intensely than the particles at the flow's periphery. Stagnation pressure at the critical point proves to be less, than the pressure at the butt's periphery (see Fig. 9.6), and there appears the tear-off zone with reversed flow. Discussed in Ref. 127 are some situations, when the reversed flow appears even by the computation of the inviscid flow. In the present case, the reason of appearance of such a flow is quite clear. As it was noted earlier, the drag of a cylindrical butt within a formed wake, possesses the weak minimum, and thus by $l/R > 2$ the drag is, practically, constant. Presented in Fig. 9.7 is the dependence of the coefficient of drag on the distance between the source and the butt, with $Q_0 = 4$. Moreover, observed is the effect of stabilization by the heat addition — $Q_0 = 25$ the lowering of a drag amounts to 55%, and with $Q_0 = 50$–57%, that is, the double increase of a heat addition leads to an additional decrease of the drag on 2% (see Fig. 9.8). Presented in Figs. 9.9 and 9.10 are the

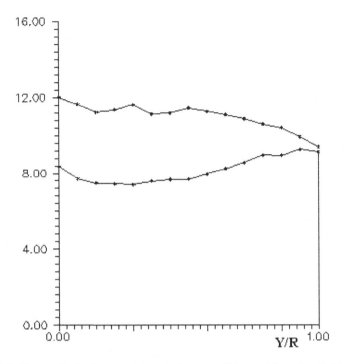

Fig. 9.6   Pressure distribution over the butt, in the cases with heater ($Q_0 = 4$) and without it.

pictures of isolines of gasdynanical parameters on some characteristical regimes (Figs. 9.11–9.17).

There was considered yet one more problem of a flow with the heat's addition — that of the flow about a cylindrical butt with a placed before it "needle of a heat". Such a problem was considered in its nonlinear formulation in Ref. 130. In the present study this problem was solved similarly to the preceding ones, that is, with the addition of heat within the "needle of a heat". The problem was solved with the parameters $M_\infty = 3.07$, $\gamma = 1.08$, $Q_\Sigma = 15$. The field of flow, obtained without a heat addition, was used as an initial approximation for the computation of a flow with the heat addition. As it is evident from Figs. 9.18–9.20, the total effect of the restructurization of flow was observed by this computation: formed were the conical shock wave and the zone of reversed circulational flow. The drag of a body was lowered from value of 1.85, down to that of 1.3. In view of a difference in the problem's formulation, the stable value of C_x is different of the result of Ref. 130. During its initial part the process has a vibrational character, while after that was observed stabilization, and such a behavior is in qualitative agreement with the data of Ref. 130.

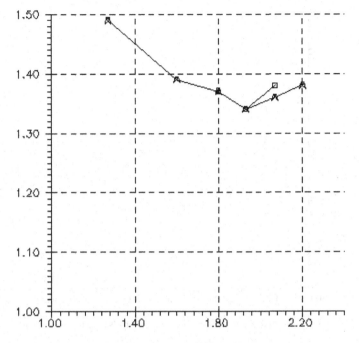

Fig. 9.7 Dependence of C_x on the distance from source to butt.

The data obtained are found to be in a good qualitative agreement with those of Ref. 126, where was considered the flow about a sphere by $M = 3$, $\gamma = 1.4$, both as concerns the effects observed, and the orders of the quantities considered.

In spite of the statistical nature of the method used, and of the first order of approximation, the most important effects and the orders of the quantities considered are found to be in agreement with the data accessible. The more accurate computation might be carried out with the use of the monotonous schemes of the increased accuracy. It seems that for the methods of DSM the computations concerning the problems of such a type appear to be not completely typical ones. The data obtained testify also the conservative quality of an algorithm. Since the accuracy in methods of DSM is determined, mainly, by the fineness of the computational network, with the growth of power of a computational technique one might hope to obtain the more accurate solutions of the gasdynamical problems of that kind. Moreover, even by the accuracy accessible one is able to solve the problems taking into account the physico-chemical processes (chemical reactions, evaporation).

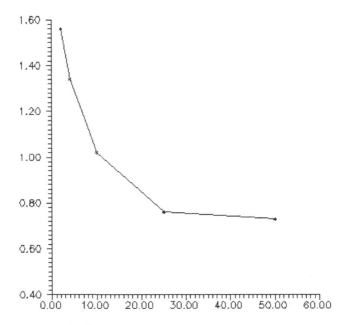

Fig. 9.8   Dependence of C_x on the source's power.

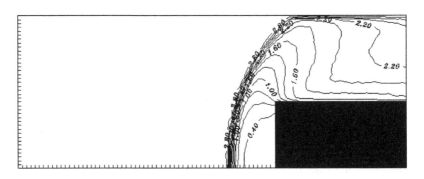

Fig. 9.9   Isolines of the Mach number ($M_\infty = 3$, $\gamma = 1.4$).

The general features of the method described consist in the fact that the structure of a flow is showed out correctly, including the subsonic areas; the numerical smoothing of the zones of rupture is carried out; the computational time needed for the main approximation is rather small, because such a computation might be realized with a small number of particles by $s \sim 0.5$–0.9. To obtain the more accurate results in a stationary case, it would be necessary for a crude approximation to

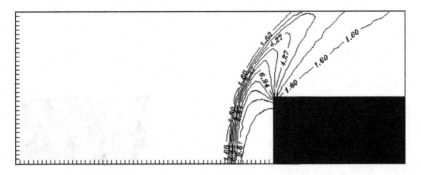

Fig. 9.10   Isolines of the pressure field ($M_\infty = 3$, $\gamma = 1.4$).

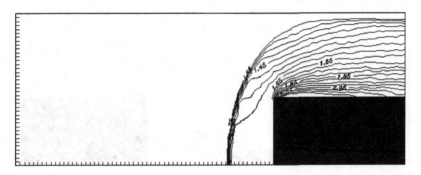

Fig. 9.11   Isolines of the field of entropy ($M_\infty = 3$, $\gamma = 1.4$).

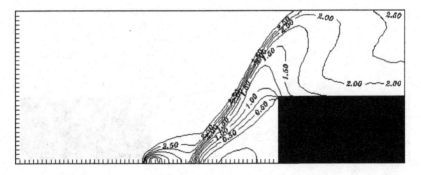

Fig. 9.12   Isolines of the Mach number ($M_\infty = 3$, $\gamma = 1.4$, $Q_0 = 4$).

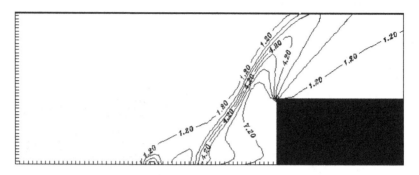

Fig. 9.13   Isolines of the pressure field ($M_\infty = 3$, $\gamma = 1.4$, $Q_0 = 4$).

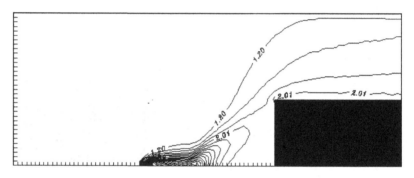

Fig. 9.14   Isolines of the field of entropy ($M_\infty = 3$, $\gamma = 1.4$, $Q_0 = 4$).

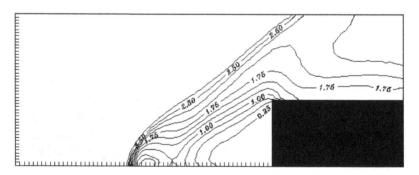

Fig. 9.15   Isolines of the Mach number ($M_\infty = 3$, $\gamma = 1.4$, $Q_0 = 50$).

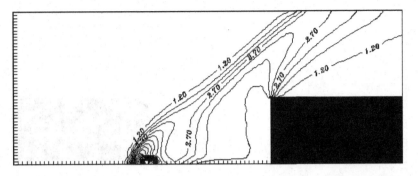

Fig. 9.16   Isolines of the pressure field ($M_\infty = 3$, $\gamma = 1.4$, $Q_0 = 50$).

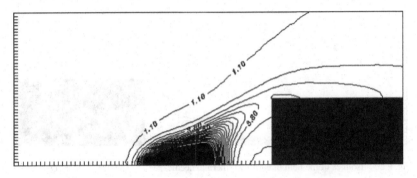

Fig. 9.17   Isolines of the field of entropy ($M_\infty = 3$, $\gamma = 1.4$, $Q_0 = 50$).

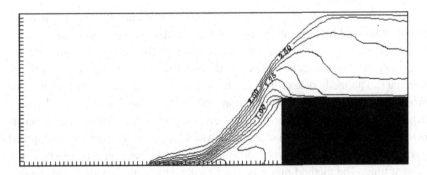

Fig. 9.18   Isolines of the Mach number ("thermal needle", $M_\infty = 3.07$, $\gamma = 1.08$, $Q_0 = 15$).

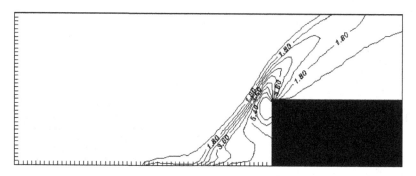

Fig. 9.19   Isolines of the pressure field ("thermal needle", $M_\infty = 3.07$, $\gamma = 1.08$, $Q_0 = 15$).

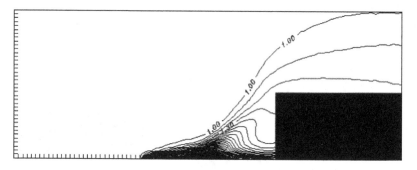

Fig. 9.20   Isolines of the field of entropy ("thermal needle", $M_\infty = 3.07$, $\gamma = 1.08$, $Q_0 = 15$).

increase, proportionally, the number of particles, and to obtain a stabilization on a new level of accuracy.

Thus, the statistical accuracy in 3–5% might be obtained with a small waste of computational time. By the use of modern computers the acceptable accuracy ($\sim 1\%$) might be also obtained sufficiently quickly. The computation of the non-stationary problem might be realized on the high-speed electronical computational machines. The method described might be effectively applied to the solution of problems of complicated configuration and of the complicated physical nature (transfer of radiation, addition of energy, evaporation, condensation, burning, magnetogasdynamical flows, and others),[6,10] where the application of the widespread gas dynamical methods meets the difficulties and is not going regularly at the present time. The method might be used for the elucidation, with the acceptable degree of accuracy, of the flow's structure, of the main effects, and of the quantitative characteristics.

Chapter 10

The General Models of Description of the Turbulent Flows

10.1 Theoretical Methods of the Description of Turbulence

There is a large amount of monographs and of scientific papers on the theoretical description of the turbulent phenomena, since this problem appears as nonwithering during more than 150 years, already. Appearing from time to time are the most bright new ideas and methods, which are inspiring the numerous researches on the getting over the extremal difficulties connected with the understanding of the essence of the problem stated. Nevertheless, the practical importance of a solution, at least of the engineering one, of this problem, gave birth to a tremendous number of semi-empirical models, which are not dealing with the essence of a problem, but just carrying out the adaptation of results for the certain set of flows possessing the practical importance. For all that, the stress is made on the description of the mean values of low-order moments: mean velocity, mean pressure, mean kinetic energy, mean concentrations of the chemical components, etc. Moreover, was developed the process of modeling, the motivation for which was reduced to the impossibility of the accurate numerical description of flows with very large Reynolds numbers.

Reached in the recent time was a considerable progress in both experimental and theoretical study of the anisotropical turbulent flows, and this fact permits to turn back to the initial problems connected with the essence of this phenomenon.[148–159] Discovered experimentally were the coherent structures, which represent the essential elements of the flows exerting the strong influence on the various physical characteristics of these flows. Thus, the flow as a whole was divided into the globally mean flow, the coherent structure and the stochastic component. The experiments were carried out, which appeared to be conductive to the revelation of the details of coherent structures. As for the stochastic component, it became theoretically connected with the so-called fractal structure of the multitude of singularities of the field of vorticity.[160–163] The singularity of the structure of a turbulent field of pulsations follows, for example, from the simple reasoning.[164]

Let us consider the equations by Navier–Stokes:

$$\partial_t \mathbf{v} + (\mathbf{v}\nabla)\mathbf{v} = -\nabla p/\rho + \nu\Delta\mathbf{v}. \tag{10.1.1}$$

By $\nu \to 0$ this equation possesses the invariancy in respect to the transformations

$$r \to \lambda r, \quad \mathbf{v} \to \lambda^h \mathbf{v}, \quad t \to \lambda^{1-h}t, \quad \lambda > 0. \tag{10.1.2}$$

For the finite values of ν the invariancy will be maintained, if

$$\nu \to \lambda^{1+h}\nu. \tag{10.1.3}$$

It is to be noted that number $R = VL/\nu$ is invariant in respect to the transformations (10.1.2) and (10.1.3). Assuming that the small-scale turbulence possesses the statistical invariancy in respect to the indicated above laws of similitude, one could choose the quantity h from the physical considerations. There is the assumption by Kolmogoroff,[141] which is formulated as: the laws of similitude in turbulence leave unchanged the flow of energy within the frame of the assumption on the locality of nonlinear interactions within the k-space. As it follows from this assumption, the rate of dissipation of the energy is invariant in respect to the transformations of similitude (10.1.3). By the definition one has $\varepsilon = \nu\langle(\nabla\mathbf{v})^2\rangle$, where $\langle\cdots\rangle$ symbolize the averaging over the ensemble. From here one has

$$\varepsilon \to \lambda^{3h-1}\varepsilon. \tag{10.1.4}$$

From the property of invariance one obtains $h = 1/3$. The Kolmogoroff's theory has a strong extension in respect to the value of a gradient of velocity $\nabla\mathbf{v}$. Let us consider the quantity

$$\Delta\mathbf{v}/\Delta x^{1/3} = [\mathbf{v}(x) - \mathbf{v}(y)]/(x - y)^{1/3}.$$

From the transformation of similitude with $h = 1/3$ it follows, that

$$\lim \Delta\mathbf{v}/\Delta x^{1/3}_{x\to y} \neq 0, \tag{10.1.5}$$

which means that the gradient of velocity proves to be the singular quantity.

Just from the initial stage of the problem's consideration, the singularity of the field of pulsations was considered by the analogy with a kinetic theory of gases, that is, the incompressible fluid was considered as an ensemble of the liquid particles — "moles". By such an approach the flow is determined by the chaotical motion of moles, every one of which possesses its own velocity and coordinate. The modification of the character of flow, as a whole, for example, the variation of the velocity field, is realized due to the turbulent mixing of the moles with different velocities of their own. Generally speaking, any one of the characteristics of a flow is the result of averaging of the proper characteristics of the moles, totality of which represents the flow under consideration. The analogy between the molar mixing within the turbulent flow and the molecular transfer in gases was used already by Boussinesq and Prandtl for the obtainment of the well-known formula for the turbulent friction. The formulae by Boussinesq looks as

$$\tau = \rho v_T \partial u / \partial y, \quad v_T = \langle lv \rangle,$$

where the quantities l and v are the accidental values of the mixing length and of the pulsational speed of a liquid particle. The formula by Prandtl looks as

$$\tau = \rho v_T \partial u / \partial y, \quad v_T = L^2 |\partial u / \partial y|,$$

where L ("mixing length") is empirical quantity. In the general case the formulae of that type are used for a realization of the connection between the two tensors: the tensor of strain and the tensor of the velocities of deformation. It is to be noted that such a connection between the tensors indicated is called "a linear" one, even if the coefficient v_T depends on the elements of velocity field.

The great contribution to the development of the theory of turbulence was done by *the cascade theory of the transfer of energy over the spectrum of turbulent pulsations*, which name means the energy transfer from the larger scales to the lesser ones. For the case of uniformity and isotropy Kolmogoroff and Obukhov gave to this theory an analytical form, using for that the theory of dimensionality and similitude. Their results were confirmed with a large degree of accuracy. From that time on, by the consideration of the flows with large Reynolds numbers the isotropical and uniform turbulence appears as the main component, in spite of the fact that might be admitted such situations, when the spectrum of energy has not yet stabilize itself. The main inference from this theory is connected with the presence of the inertial area of the spectrum over the wave numbers $k : 1/L \ll k \ll (1/L) R^{3/4}$, within which the viscous effects of the energy dissipation are inessential, and due to that fact the spectral density of energy varies in dependence on the wave number according to the law "$-5/3$". The inclusion of this element into the fluid dynamics leads to the appearance of models possessing a certain phenomenological connection between the tensor of strains and that of velocity deformations, and this connection realized through a number of additional equations for the variable quantities of the type of the turbulent energy, of the rate of dissipation, etc. (like, for example, the $K - \bar{\varepsilon}$ model). However, as it was shown in practice, the models of such a type have a rather narrow area of application. Simultaneously with the change of the application area are changing the constants, entering the corresponding equations, and these constants should be determined anew, experimentally. Moreover, for the flows of the type of a boundary layer appeared the difficulties connected with satisfaction of the boundary conditions at the solid surfaces.

The more logical, in our opinion, direction of the construction of models for the flows with large Reynolds numbers is connected with the so-called *sub-network modeling*, of which the sense is connected with the intention to leave in the hydrodynamical equations only those scales, which surpass the sizes of computational network (permitted scales). This measure decreases the number of the degrees of freedom down to some reasonable value and permits to use the modern computer technique for the determination of the average fields of flow. The size of a computational network is chosen in such a way, that the corresponding to it wave number appears to be within the inertial area, and introduced is a certain connection between the tensor of strains and the elements of the flowfield. Thus, for example, in the model of Smagorinski is introduced the linear connection between the tensor of strains and that of the rates of deformation. The viscosity coefficient is changed for the coefficient of turbulent viscosity, which is determined from the averaging of sub-network pulsations, that is those pulsations, the size of which is less than that of network. Added to the initial equations might be several additional equations, as, for example, the equation for the sub-network kinetical energy, and so on. The equations are solved temporally, in respect of the permitted variables, and in this process the pulsations with sub-network frequencies are filtered out with the help of one or another filter, and the remained quantity is averaged over time. This property represents the main distinction of the models described from those of Boussinesq or Prandtl, which might be used in the stationary formulations, too.

However, as it was shown in practice, the strongly anisotropic flows, like the flow in the boundary layer or in the mixing layer, are not embraced by the theories of that type, which might lead to the spoiled profile of velocity or to another similar effects. As it is shown by the last experimental achievements, the models of that type are not able to catch some effects observed in the real flows. After the scrupulous analysis it was found that the sub-network models ought to contain in them the effects of energetical transfer over the spectrum in the inertial area, including the reversed dissipation of energy, as well as its redistribution between the normal components of the strain tensor. These effects appear as a consequence of the nonlinear interactions and anisotrophy. Used in the procedure of modeling proposed here is the phenomenological approach, in which the sub-network strains are connected with the permitted velocity gradients by way of nonlinear relations. Presented in the survey is the nonlinear model containing the effects of anisotropy within the boundary-layer flows. The results obtained due to the use of this nonlinear model during the large-scale modeling of a neutral, shift-like boundary layer within the atmosphere, demonstrate the considerable improvement in the forecast of mean values in their comparison with linear models of the type of Smagorinski model. These results indicate, also, at the strong influence of a model on the structure of flow.

Discussed above were the methods of sub-network modeling, but besides of those, the wide circulation obtained the methods of statistical modeling of the turbulent flows.[173–177] In these methods an attempt is made to construct, on the phenomenological basis, the equation for the density of probability of fluctuations of the velocity field (and of the field of other parameters), which is solved afterwards with the help of Monte Carlo methods. Such an approach permits to compute not only the middle moments of the lowest order, but also the more delicate statistical characteristics.

As at one practical achievements of these approaches one might indicate at the numerical solution of such problems, as the turbulent wake behind the cylinder, the dissipation of a turbulent spot, the profile of a turbulent boundary layer (TBL), the flow about reversed step, and so on.

Parallelly to the indicated resultative approaches to the description of dynamics of turbulence are developing the theoretical studies, in which on the basis of Navier–Stokes equations are made the attempts either to find the statistical solution of the problem[163,165,166] (the problem of closing, equations in functional derivatives), or to use the methods of dynamical systems (multifractal structure of the field of vorticity, wavelet analysis, i.e. the fractal transformation of curtail),[160–162] or yet to use the previously justified themselves by the study of critical phenomena renormgroup-like applications of the theoretico-physical *asymptotical methods*, developed in respect to the description of the dynamical systems with the infinite number of *the degrees of freedom and with the stimulation of the continuous spectrum of scales* (see Shirkov, Teodorovich[167–170]). The details of this are extremely cumbersome. However, the essential features of some versions of it might be elucidated on the example of method by Gauss[171] for the calculation of the elliptical integral (in renormgroup method (RNG) methods are also calculated the integrals for the determination of values, averaged over the ensemble, though these integrals prove to be, in general case, the continual ones) with the help of the arithmetico-geometrical averaging. Let us assume that it is necessary to calculate the integral looking as

$$I = \int_0^{2\pi} \frac{d\varphi}{2\pi\sqrt{m^2 \cos^2\varphi + n^2 \sin^2\varphi}}.$$

For the realization of calculation of this integral is made a transformation, which leaves both external appearance and value of this expression unchanged, that is, the new expression might be written down as

$$I = \int_0^{2\pi} \frac{d\varphi'}{2\pi\sqrt{m'^2 \cos^2\varphi' + n'^2 \sin^2\varphi'}},$$

while m' and n' are determined by the expressions

$$m' = \frac{m+n}{2}, \quad n' = \sqrt{mn},$$

that is, in the recurrent form. As it was proved by Gauss, the limit for m' and n' in these recurrent interrelations exists, and m' and n' coincide. After all that the integral I, which is looked for, is easily expressed in terms of the limit mentioned. The reduction of a problem to the recursive relations proves to be the basis of the numerous RNG approaches. Up to the late times the works by the last group were considered as excessively abstract and possessing the purely methodological value. However, the paper by Orszag and Jachot has shown in 1986 that it is not true, though excited the storm of negative comments from the theoreticians. Developed in this paper was the version of the RNG for the description of hydrodynamical turbulence. This procedure, in which are used the dynamical similitude and invariance along with the iterational methods of the theory of perturbations, permits to calculate the coefficients of transfer and to determine the equation of transfer for the large-scale (slow) modes. The RNG theory does not contain any parameters, determined experimentally, and gives the numerical values for the constants of turbulent flows: Kolmogoroff's constant for the energy spectrum within the inertial interval, $C_K = 1.617$; turbulent Prandtl number, $Pr_t = 0.7179$; constant by Bachelor, $Ba = 1.161$; the coefficient of asymmetry, $\bar{S}_3 = 0.4878$. Obtained is the differential $K - \bar{\varepsilon}$ model, which in the area of large Reynolds numbers gives the algebraical relation for $\nu = 0.0837 K^2/\bar{\varepsilon}$, while for the uniform and isotropical turbulence the value K is decaying like $K = O(t^{-1.307})$ with the Karman's constant $k = 0.372$. The differential equations of transfer, which are based on the differential connections between $K, \bar{\varepsilon}$, and ν, are obtained in such a form, which does not lead to a disconvergence by $K \to 0$, for the finite value of $\bar{\varepsilon}$. It was assumed that this last model would be especially useful by the description of flows near the solid walls.

The authors mentioned have obtained the model of the type of Smagorinski model, which, as it was already indicated earlier, does not contain the number of such phenomena, without which it is impossible to give the correct description of the turbulent motion of a fluid. Nevertheless, the possibility of using the mighty theoretical apparatus for the inference of equations for the slow motions of a fluid was taken as an arms by the numerous authors. The further development of these methods has led to a real possibility for the analytical inference of the equations of dynamics of the long-wave component of the turbulent flows, and of the coefficients contained in these equation, without the addressing to the experimental data. One should note that the difference in approaches by different authors leads to a rather insignificant difference in their results!

Considered in the present monograph are the most promising methods of the inference of the equations of motion of the incompressible turbulent fluid, as well as some questions, arising in this connection from the point of view of the traditional asymptotical methods of mechanics.

In this connection it would be interesting to indicate at the results of Ref. 172 and some similar papers, in which, on the basis of one of the versions of the RNG, of the so-called *method of recursive renormgroup*, were theoretically obtained the equations of motion of the fluid with a nonlinear connection between the tensor of strains and that of the ratios of deformations, and the results of the corresponding models prove to be strongly different from the results of models of the Smagorinski type.

Presented in the reference cited was the alternative version of RNG theory (so-called recursive RNG, or r-RNG, theory), connected with sub-network modeling of turbulence, which is independent of the order of the execution of the sub-network averaging. In the explicit way are considered the relevant approximation, the perturbative ordering and the process of averaging. In particular, it is shown that there appears the nonlinearity of high order appearing in the r-RNG equations by Navier–Stokes, which does not surpass the third order at the desirable level of the perturbative disturbances. Moreover, these terms with triple products of the velocity components appear just in the same order as the vertical viscosity, which is generated during the belonging to the RNG-procedure process of the exclusion of sub-network scales. These nonlinearities of the third order play, also, the important role in the equation of the energy balance, which occurs in connection with the corresponding process of the energy transfer and appears in the analytical formulation of vortical viscosity which is coordinated with the same vortical viscosity in the theories of closing and with the results of numerical solutions of the initial equations. This fact is confirmed also with the help of a direct analysis both by the method of modeling of large vortices, and by the analysis of the data on velocity field obtained by way of the direct numerical modeling. Moreover, it was shown that the induced by RNG triple nonlinearities lead to the appearance of the reversed flow of energy, reflected from the smaller scales in the direction of larger scales, and this phenomenon is in agreement with the last achievements of the theory of closing and with the results of numerical modeling.

The equations obtained were used later, for example, for the description of the accompanied by separation flow about the reversed step. The computational results showed that the mean characteristics of a flow, which are determined from the unsteady solution of the equations, appear to be in much better correspondence to the experimental data, as compared to the other models.

10.2　Coherent Structures in the Turbulent Boundary Layer (see Khlopkov, Zharov, Gorelov[205])

10.2.1　*History of a problem*

At the present time is obtained a tremendous quantity of data on the dynamics of a turbulent motion of incompressible fluid within the boundary layer at the plate, discovered in which are the structural elements of the interaction of the surface in flow with the flow itself. The experimental data are obtained by the different researchers, using the different methods, and these data reveal some general features of that interaction. Now is a proper time for the intensive theoretical comprehension of these results and for the construction on their basis of the global theories of turbulent motion. Listed below, in the most concise form, are the main results of the experimental research, and, in our opinion, these results might be actively used by the construction of mathematical models.

The decomposition of the nonstationary motion was, for the first time, proposed to be done by Reynolds in the year 1894.[178] The quantities, describing the flow, were divided into the mean part and fluctuating one, and as a result of that, after the substitution into equations by Navier–Stokes was obtained the set of equations of the form, identical to that of the initial set with an exception of the terms with convective strains, generated by the averaged products of the fluctuations of velocities. In order to close the set of equations, it is necessary to have yet one additional relation between the convective strains and the field of mean velocity. Up to the recent times, the numerous theoretical and experimental studies were concentrated on the search of the connections, which might be applied to the ever growing diversity of the mean flows with the hope to find a certain universal relation for the "turbulent viscosity". The hope for discovery of the universal model of turbulence was gradually changed for the growing confidence in such an idea, that the formulation of the adequate theory demands for the considerably better understanding of the physics of a turbulent flows.

For the researcher of 1920–1930 years turbulence appeared as the essentially stochastical phenomenon possessing the well defined and reproducible mean value, imposed on which was the stochastically fluctuating velocity field. The motion was characterized by the wide area of scales, which were limited only by the full sizes of a flow. This picture of the stochastically interacted elements of flow with a number of different scales led to the appearance of the semi-empirical theories by Prandtl[179] and by Taylor,[180] in which the convective strains were connected with a mean flow with the help of the effective vortical viscosity (which was introduced by Boussinesq in 1877), or with that of a mixture length.[181]

The consideration of evolution of the spectrum of a uniform and isotropic turbulence has led to the important observation, that the energy-possessing structures of the isotropic turbulence do not directly depend on the value of viscosity of the field. It proved that by the sufficiently large Reynolds numbers, when the fluid is wittingly turbulent, the energy-possessing structures prove to be similar for all the values of Reynolds number. It was noted, in addition, that if the Reynolds number is sufficiently large, then the zone of dissipation and zone of the generation of turbulent energy prove to be strongly divided in the space of the wave numbers. In this case the small-scale motion is in the state of a local isotropic equilibrium.

The new element was added to the physical picture of turbulence by Corrsin[182] and by Townsend,[183] who have shown that the external boundaries of the turbulent shear flows, especially in the jets and wakes, are just only in the state of intermittent turbulence. In the middle 1950s of the 20th century the physical notions were successfully developed, up to the state, which might be characterized by the depicted below picture (see Townsend[185]), which is connecting the laminar sub-layer by Corrsin and Kistler[184] with the field of a turbulent fluid having almost uniform intensity. The dynamical characteristics of this field are similar to the characteristics of the isotropic turbulence. The turbulent fluid is set into motion by the slow convective pile-up of the totality of large vorticies, of which the sizes are comparable with the width of a flow and much larger, than the scale of those vorticies, which contain the larger share of a turbulent energy (see Fig. 10.1).

Within the area of flow, which is external in respect to a surface of the turbulent division, the motion is unsteady. Induced within that area, by the motion of a fluid near the boundary, is the potential flow without vorticity.

Townsend considered the large vortices and small-scale turbulence as the main feature of a double structure and stressed the important role of the large vortices

Fig. 10.1   Boundary layer at the flat plate (from the album by VanDyke[207]).

in the process of transfer. He stressed also that the large vortices should accept the quasi-deterministical form and tried to draw the picture of a large-scale vortical motion, sufficiently realistical for that period of time. Any attempt to create such a picture acquired usually the form of conclusions, based on the averaging over the large interval of time of the spatial correlational tensor[211]:

$$R_{ij}(\bar{x}, \bar{\xi}) = \langle \bar{u}_i(\bar{x}, t), \quad \bar{u}_i(\bar{x} + \bar{\xi}, t) \rangle,$$

which is measured in the Eulerian system of counting. The extensive correlational measurements[186–189] in various turbulent flows have revealed the number of characteristical formations, which might be interpreted as the coherent vortical structures.

In the early 1960s of the 20th century were conducted the experiments, which began to modify the estimative views on turbulence. In the last 20 years of the research on turbulence appeared the confidence in that the characteristics of a transfer of momentum, energy, etc., of the turbulent shear flows are determined by the large-scale vortical motion, which is not in an accidental form, but is a deterministical one. The form, intensity, and scale of these organized motions is varied from one version of flow to another, and simultaneously are modified the methods of their determination.

10.2.2 *The structure of an averaged flow*

The earliest observations of the organized motion were done in the TBL, which was flowing along the wall, where the flow is the most complicated. That is stipulated by the fact that TBL is just the flow, which always attracted the strongest attention due to its technical importance, and it was extremely desirable to reveal its structure in the first turn.

The mean profile of velocity within the TBL might be divided in three parts (see also Ref. 190):

the viscous sub-layer: $0 \leq y^+ \leq 7, \quad u^+ = y^+$,
the buffer sub-layer: $7 \leq y^+ \leq 30$,
the logarithmical and external layers: $30 \leq y^+ \leq \delta^+$.

There are several empirical formulae for the description of the mean component of the longitudinal velocity[206] $u^+(y^+, y/\delta)$, with δ as a thickness of a boundary layer, which are expressed through the set of variables

$$y^+ = \frac{yu^*}{\nu}, \quad \delta^+ = \frac{\delta u^*}{\nu}, \quad u^+ = \frac{u}{u^*}, \quad u^* = \sqrt{\frac{\tau_w}{\rho}}, \quad \tau_w = \mu \frac{\partial u}{\partial y}\bigg|_{y=0}.$$

The first remarkable property of the TBL is the universality of its behavior near the wall. Independently of the value of a pressure gradient, of the wall's roughness, or of the Reynolds number, observed is the logarithmical dependence of the velocity u on the coordinate. Moreover, the summation of the speed of generation of a turbulent energy over the whole thickness of a boundary layer leads to a result, that the first 5% of a boundary layer bring more than a half part of the turbulent energy generated. This result appeared to be the first stimulus for the early studies and remains to be the same for the larger part of studies on the structure of a TBL, which are carried out up to the present moment.

It is to be noted that considered in Ref. 192 is the possibility of existence of the another structure of a TBL (the power-type behavior near the wall), which transforms itself into the described above form only when the Reynolds number is tending to infinity.

10.2.3 *The internal area*

Beginning with the late 1950s of the last century in Stanford were initiated the series of experiments with the use of the flow's visualization, directed to the research of a TBL. These efforts were concentrated in the work by Kline and others.[191] There was revealed a number of new properties of the flow of TBL near a wall. Observed during the experiments was the interaction of streaks, the high-speed and low-speed areas of the by-wall flow, with the external parts of that flow, through the sequence of the four events: the slow uplifting, the ascent, the sudden oscillations, and the collapse. The sequence consisted of three events, from the ascent till the collapse, the researches named as "bursting" (Fig. 10.2).

In addition to that they have found that the favorable gradient of a pressure, $\partial p/\partial x < 0$, aspires to a diminishment of the frequency of bursting, while the infavorable tends to increase this frequency. It was confirmed that the bursting is playing the decisive role in the generation of a turbulent energy, that its influence prevails in the processes of transfer between the inner and the external areas and, thus, that the bursting plays an important role in the determination of structure of the layer as a whole. Kline and others[191] proves to be able to estimate the spatial scales of motion, connected with the streaks and bursts. From the data of visual observations they came to a conclusion that the mean cross-sectional distance between the streaks (that is, the distance, equal to the length of one wave) for the smooth surface and by the arbitrary pressure gradient was equal to, approximately, $\lambda_z^+ = \lambda_z u^*/\nu = 100$ (see Fig. 10.3).

Being collected together, these initial observations induced the further research on the TBL, which has demonstrated the important connection between the

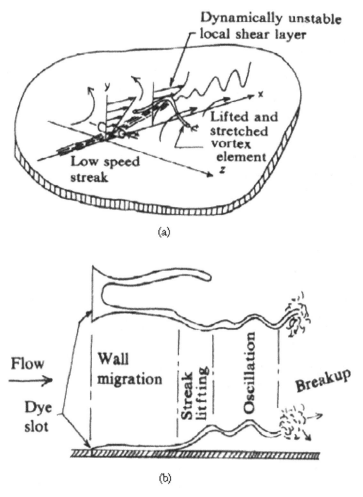

(a)

(b)

Fig. 10.2 Schematic presentation of the unsteady structure of a turbulent boundary layer. (a) Mechanics of the streak's breakup,[191] (b) succession of the events in (a).

quasi-deterministical, repeated and nonstationary motions of a fluid, the generation of turbulence and the support of the mean turbulent flow.

After the check of data within wide range of the Reynolds number, $600 < Re_\delta < 9000$, Rao and others[193] have shown that even within the layer by a wall the mean period of a burst is expressed in terms of the external variables (U_∞, δ), with the mean length of time between the bursts, equal to

$$\frac{U_\infty T}{\delta} = 6.$$

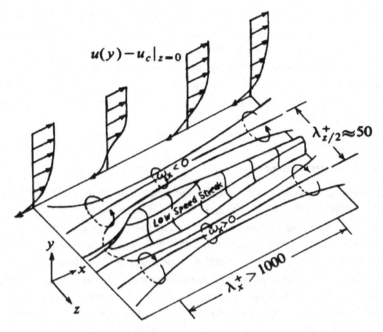

Fig. 10.3 The longitudinal vortices rotating to meet each other and streaks as the result of these rotations.[195]

Moreover, they have found that the mean frequency of the bursts did not essentially vary with the change of a distance from the wall. The authors have assumed also that the mixed version of nondimensionalizing for the internal variables for the cross-sectional spatial scales, and for the external ones for the time between the bursts, which leads to the appearance of a parameter $U_\infty u^*/F\delta^*\nu$ (where δ^* is the thickness of displacement, F — the frequency of the burst for a unit span), as to a quantity, practically independent of the Reynolds number.

As it was found in Ref. 194, the ejection and the splashes were observed independently of the degree of a surface's roughness. The by-wall flow within the boundary layer over the completely rough surface should be strongly different from the flow over the smooth surface. It is evident, however, that the main structure of an organized motion does not undergoing any essential variations. The similar data, mostly indirect ones, exist also in respect to the surfaces with a slight blow-in or blow-out (see, for example, Ref. 28), to the wavy surfaces or to the surfaces, oscillating in the cross-sectional direction.

Blackwelder and Eckelmann[195] have fullfilled the sufficiently detailed research of the structure of by-wall streaks, directed at the confirmation of the presence of a longitudinal and a cross-sectional vorticity at the wall (see Fig. 10.3). They have found that the intensity of longitudinal vortices is in one order less than the mean cross-sectional vorticity. They singled out the low-speed streak, which was observed by Kline and others as the area of an accumulation between the longitudinal vortices, where the vertical component of a secondary motion is directed away from the wall. They have found also, that the length of longitudinal vortices is equal to $\Delta x^+ \sim 1000$.

10.2.4   *The external area*

(Here as a general rule is adopted the terminology, according to which the "by-wall layer" is referred to $0 < y^+ < 100$ and contains in itself the viscous sub-layer and the part of a logarithmical area. All the rest is included into the external area).

Kovasznay and others[196] have used the conditionally averaged autocorrelational functions of the several variables, which describe the flow, with the aim of drawing the three-dimensional chart of an external structure (see Fig. 10.4). They discovered that vorticity experiences the breaks when passing across the boundary separating the turbulent nonuniformities from the external flow, though the velocity is at the same time continuous. In addition, they noted that between the forward and backward parts of the turbulent formations exists the perceptible difference, revealed in the external flow. They have confirmed that if the flow is considered in the system of counting, which moves with a mean velocity of the boundary of division, then the fluid moves in direction of a stagnation point, which is located behind the nonuniformity, at the distance approximately equal to $y/\delta = 0.8$. This result was obtained also during the latest investigations of the organized structures within the turbulent flow, and it means that the backward part of the turbulent formation appears to be the most active one, and its activity is connected with a saddle point in the moving system of counting.

Brown and Thomas[199] connected the shift at the wall with the velocity across the layer. As it was shown by these authors, the shift of velocity near the wall had the slowly varying part and high-frequency part, and that both these parts were interconnected. In this process was determined the line of a maximal correlation, which is inclined in respect of a wall under the angle of 18° in longitudinal direction, and it was assumed that such a behavior is a sequence of the fact, that the organized structure is inclined at the sharp angle in respect of a wall, and that at the wall is created the characteristical perturbation of a strain by the motion of that organized structure along the plate with the velocity of $0.8\,U_\infty$. At the backward part of the surface of division was discovered a sharp jump of velocity. Falco[197] has combined the visual and thermoanemometrical measurements in the outer part of the surface

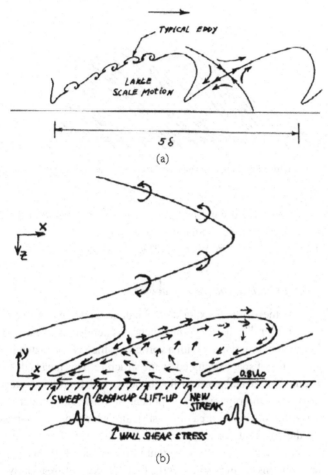

Fig. 10.4 Model of the external structure of a turbulent boundary layer.[196–198]

of division in the external structure, which was connected with the motion, created due to the generation of Reynolds strains by the small-scale vortices within the external area, with the characteristical scale, equal to $(100–200)\,\nu/u^*$. Drawn in Refs. 197, 198 was the scheme of an organized structure, quite similar to the described one.

Head and Bandyopadhay[200] have used the flow's visualization and thermoanemometrical measurements for the drawing of a quite different picture. These authors assumed, that for the values $Re_\delta = 1000–7000$ the most characteristical feature of a boundary layer is the existence not of a large-scale coherent motion,

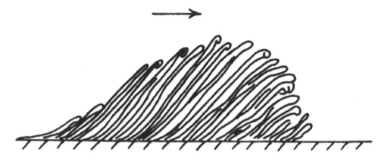

Fig. 10.5   Model of a structure of turbulent boundary layer on the basis of strongly stretched horseshoe vortices.

but of the structures, formed with the help of accidental merging of the elements having a small size in the longitudinal direction, but stretched out along the lines, inclined at the angle of 40° in respect of surface (see Fig. 10.5).

10.2.5   *The definition of a coherent structure*

In accordance with the available experimental data it would be possible to provide the following definition[201]: the coherent structure is defined as a multi-connected, large-scaled, turbulent liquid mass with a vorticity, which is correlated in its phase within the whole area of space occupied by the structure. This means that the latent three-dimensional, accidental vortical fluctuations, which are characterizing the turbulence, appear as the organized component of vorticity, correlated in phase (i.e. coherent) within the whole area of the structure's location. The largest spatial size, at which the coherent vorticity exists, is just the size of coherent structure. Thus, the turbulence consists of the coherent motions and of the accidental ones; the last of these motions are imposed over the first of them and are usually stretched far away from the boundaries of the coherent structure.

10.2.6   *The triple expansion*

In the presence of a coherent structure one might consider the instantaneous field of flow, which is, actually, a sum of the three components[202]: the component, independent of time, the coherent component, and the noncoherent turbulence. This means that for any instantaneous variables one has

$$f(x, y, z, t) = F(x, y, z) + f_c(x, y, z, t) + f_r(x, y, z, t),$$

where $F = \bar{f}$, $\bar{f}$ — the quantity, averaged in time in respect of f, f_c — the quantity, averaged in phase, f_r — the accidental component. Thus, at the arbitrarily chosen point, by the condition that the phase of a structure is also known, the three versions of a field might be determined.

10.2.7   *Conclusions*

The results listed do not leave any doubt concerning the existence of the coherent dynamical formations, with the help of which is realized the generation of vorticity directed from the surface into the flow.

Nevertheless, this totality of the experimental data possesses certain deficiencies. For example, there is a lack of certitude about the question, which processes contribute to the main part of the tangential strains at the wall. A certain part of the experimental data states that this is a result of the single bursting. In the other experiments, the generation of tangential strains at the wall is connected with the large gradients in the bottom of a characteristical vortex, which appears as a result of the multiple action of the burstings. In this connection one should note also, that there exists an uncertitude concerning the determination of the bursting's frequency, because this frequency might be determined either with the help of a single device, or as a frequency of the bursting-type events, counted for a unit length of the device's comb (in the survey monograph[205] is presented more detailed description of these and many other, similar phenomena, occurring in TBL).

And, after all, the totality of the results presented is, in our opinion, sufficient for the realization, on their basis, of the effort of formulation of such mathematical models, which would describe the process of the vorticity generation from the solid surface of a body in flow. Earlier, within the frames of vorti-potential flows were formulated the "static" models of dynamical structures, which have taken into account the presence of longitudinal vortices[149] (see Fig. 10.6), as well as a double structure of the totality of hairpin-like vortices[201] (see Fig. 10.7), with the help of which one succeeded in the very accurate description the profile of a mean longitudinal velocity.

Presented in Ref. 209 (see Fig. 10.8) are considerably more delicate details of the vertical structure. Nevertheless, the modern technique demands to carry out the further generalization of the experimental data and, in particular, the creation of nonstationary models of the dynamical structures describing the vorti-wavy processes in TBL (see Fig. 10.9[210]). In this direction are executed certain efforts (see, for example, Ref. 203, where proposed is the algorithm of the explicit singling out of the coherent structures within the frames of a triple expansion of the velocity field).

Thus, appearing are the reasons to believe that the experimental data, presented above, will permit to formulate, as well, some other sapient models, which will be

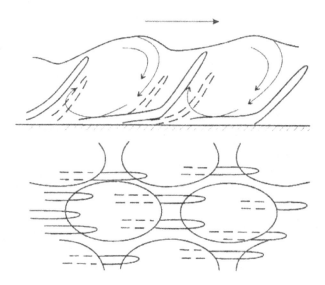

Fig. 10.6   Turbulent boundary layer according to Fiedler.[208]

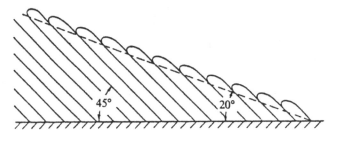

Fig. 10.7   Double structure.

able to reveal in a full measure the possibilities of control in respect to the above described phenomena.

10.3   The Description of Turbulence with the Help of a Model of the Three-Wave Resonance

Let us consider the equations by Reynolds, for the mean values U and V, and for pulsations u, v, w, p, in the approximation of a boundary layer. Let d be a characteristical cross-sectional scale of the flow, L — longitudinal scale. By the estimation of variables in the equations of initial form are compared the characteristics of the type of force, and for that reason one should take that $d \approx \delta^{**}$, δ^{**} — the thickness of a momentum's loss. Thus we have $\varepsilon^2 = d/L \ll 1$. Practically, $\varepsilon^2 \ll 10^{-2}$.

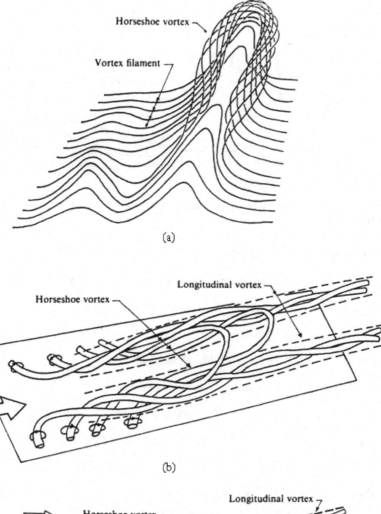

Horseshoe vortex

Vortex filament

(a)

Longitudinal vortex

Horseshoe vortex

lain flow
irection

(b)

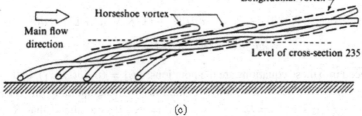

Longitudinal vortex

Horseshoe vortex

Main flow
direction

Level of cross-section 235

(c)

Fig. 10.8   One of the last schematic models of vortical structure.

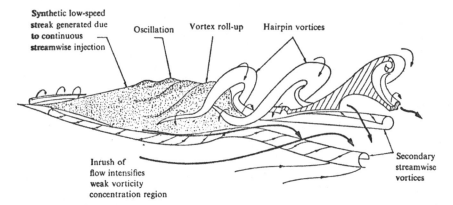

Fig. 10.9 Schematic presentation of the unsteady process of a low-speed streak with formation of hairpin vortices.

Let us introduce the notations

$$\bar{U} = U/U_\infty, \quad \varepsilon^2 \bar{V} = V/V_\infty, \quad \bar{u}_i = u_i/U_\infty,$$

$$i = 12, \ldots, 3, \quad \varepsilon\bar{p} = p/\rho U_\infty^2, \quad \bar{x}_i = x_i/d,$$

$$\bar{t} = tU_\infty/d, \quad X = x/L, \quad T = tU_\infty/L,$$

$$\{u_i\} = (u, v, w), \{x_i\} = (x, y, z).$$

Then the equations for $\bar{U}$ and $\bar{V}$ are written in the form

$$\partial\bar{U}/\partial T + \bar{U}\partial\bar{U}/\partial X + \bar{V}\partial\bar{U}/\partial\bar{y}$$

$$= -\partial\langle\overline{uv}\rangle/\partial\bar{y} + (1/\varepsilon^2 R)\partial^2\bar{U}/\partial\bar{y}^2,$$

$$\partial\bar{U}/\partial X + \partial\bar{V}/\partial\bar{y} = 0, \quad R = U_\infty d/v,$$

and for $\bar{u}_i, \bar{p}, i = 1, 2, 3$, one obtains

$$\partial\bar{u}_i/\partial\bar{t} + \bar{U}\partial\bar{u}_i/\partial\bar{x}_1 + f_i = -\partial\bar{p}/\partial\bar{x}_i + (1/R)\nabla^2\bar{u}_i + \varepsilon T_i + o(\varepsilon^2),$$

$$\partial\bar{u}_i/\partial\bar{x}_i = 0, \quad R = U_\infty\delta^{**}/v.$$

Introduced into these equations are the notations: $\{f_i\} = (u_2(\partial\bar{U}/\partial\bar{x}_1), 0, 0), T_i = \partial(\langle\bar{u}_i\bar{u}_j\rangle - \bar{u}_i\bar{u}_j)/\partial\bar{x}_j$. For the boundary conditions for these equations one is taking: $\bar{u}_i = 0, i = 1, 2, 3, \bar{y} = 0, \infty; \bar{p} = 0, \bar{y} = \infty$. It is assumed that $R \gg 1$, $\varepsilon^2 R \gg 1$.

This set is transformed into equations for the vertical components of velocity and vorticity, by the assumption of the absence of resonance for the

Tollmien–Schlichting waves (T–S) with the waves by Squire. Further on, the equation for the vertical component of vorticity will be, for the sake of simplicity, dropped out. As soon as the vertical component of velocity is exposed to the expansion in the terms of the lowest discrete spectrum of the T–S waves, one will obtain

$$\frac{\partial a_{\bar{k}}}{\partial t} + i(\bar{\omega}(X_0, \bar{k}) - \dot{R}_0 \bar{k}) a_{\bar{k}}$$

$$- \varepsilon^2 \frac{\partial}{\partial x} \left(\frac{\partial \bar{\omega}(X_0, \bar{k})}{\partial_0^X} a_{\bar{k}} \right) = \varepsilon \int H_{\bar{k}\bar{k}'} a_{\bar{k}} a_{\bar{k}''} d\bar{k}' + \cdots ,$$

$$\bar{k}'' = \bar{k} - \bar{k}',$$

where $k = \bar{k} = (\alpha, \beta)$, $\bar{\omega}(X_0, k)$, $\bar{R}_0 = (X_0, Z_0)$ are, correspondingly, dimensionless wave vector, the own numbers of the Orr–Sommerfeld equation (O–S), (X_0, Z_0) — "the center of mass" for the coordinates of a wave packet. The amplitudes α_k are connected with the vertical velocities of perturbation $v(x, y, z)$ in the following way:

$$v_{\bar{k}}(y) = a_{\bar{k}} \frac{v_0(y)}{N} + \cdots ,$$

$$N = -\int_0^\infty \left(\frac{d\hat{v}_0(y)}{dy} \frac{dv_0(y)}{dy} + k^2 \hat{v}_0(y) v_0(y) \right) dy,$$

where $v_0(y)$ is the lowest mode of T–S waves, $\hat{v}_0(y)$ — the own function, $v_k(y)$ — the Fourier–image of $v(x, y, z)$.

If the proper assumptions are made, then in this case for "the density" of the wave packets, $I(k) = \langle a_k^* a_{k\text{©}} \rangle / \delta(\bar{k} - \bar{k}\text{©})$, one obtains the equation of transfer:

$$\partial I(\bar{k})/\partial T + g(\bar{k}, X)\partial I(\bar{k})/\partial X - h(\bar{k}, X)\partial I(\bar{k})/\partial \bar{k} - 2\bar{\omega}^I I(\bar{k}) = J_c,$$

$$g(\bar{k}, X) = \partial \bar{\omega}^R(\bar{k}, X)/\partial \alpha, h(\bar{k}, X) = \partial \bar{\omega}^R(\bar{k}, X)/\partial X.$$

This equation is reminding the kinetic one, and it describes the birth of quasi-particles in the area of instability of the lowest mode of T–S waves, their transference with a group velocity $g(k, X)$ under the action of a "force" $(-h(k, X))$, as well as the disintegration and merging of particles due to the three-wave resonance, which is described by the term J_c. In the elementary case the integral of collisions of such a type has a form

$$J_c = \int d\bar{k}_1 P(\bar{k}, \bar{k}_1) \delta(\bar{\omega}^R(\bar{k}) - \bar{\omega}^R(\bar{k}_1)$$

$$-\bar{\omega}^R(\bar{k}_2))(I(\bar{k}_1)I(\bar{k}_2) - I(\bar{k})I(\bar{k}_1) - I(\bar{k})I(\bar{k}_2)),$$

$$\bar{k} = \bar{k}_1 + \bar{k}_2.$$

For the study of equation for the wave packet $I(k)$, which is similar to the kinetic one, might be applied the method of direct statistical modeling, which was successfully used by the modeling of those phenomena, which are described by the kinetic equations.

10.4 The Fluidical Model of the Description of Turbulence (Belotserkovskii, Yanitskii)

Rather long time ago Prandtl turned his attention to the fact, that there is an analogy between the rarefied gas and the turbulent fluid. As an extension of the application of kinetic models to the continuous medium, in Refs. 16 and 21 was made an attempt of description of the turbulent phenomena. In particular, studied was the example of the dissipation of a turbulent spot. Here, like it was in dynamics of a rarefied gas, the problem is solved at the level of a distribution function. Now, however, as an argument appears not the molecular velocity ξ, but the pulsations of the velocity of a liquid particle v. In this model, each of the particles in a cell possesses the new quality (see Table 1).

The liquid particle, just as it was earlier is characterized by the physical coordinates and by velocity. For this distribution function proposed is the model of a kinetic equation, similar to the model equation in the dynamics of rarefied gases.

The problem of a dissipation of the turbulent spot.

By the consideration of this problem the principal aim consisted in the preservation of the main principles of direct statistical modeling.

Table 1. Description of the medium with the help of distribution function.

Rarefied gas dynamics	Turbulence
Particles	
Molecules	Liquid particles
r_i, coordinates of molecules	x_i, coordinates of particles
c_i, velocities of molecules	v_i, velocities of pulsations
Distribution function	
For the molecules	For the liquid particles
$f = f(t, r, c)$	$F = f(t, x, v)$
$\int f dc = \rho$, density	$\int f dv = 1$, standardization
Moments	
$\frac{1}{\rho} \int c f dc = u$, macroscopic velocity,	$\int v f dv = u$, mean velocity
$(c - u)$, thermal velocity	$(v - u)$, fluctuations

For the description of a turbulence is used the kinetic equation by Onufriev–Lundgren:

$$\frac{\partial f}{\partial t} + v\frac{\partial f}{\partial x} - \frac{1}{2\tau_1}\frac{\partial}{\partial v}(v'f) = \frac{f_M - f}{\tau_2},$$

where $f_M = \left(\frac{3}{4\pi E}\right)^{3/2} \exp\left[-\frac{3v'^2}{4E}\right]$ corresponds to the normal law, and E is the turbulent density of energy.

This kinetic equation is similar to the model equation by Krook, which was described in Chap. 1.

Here the scheme of modeling is built up in accordance with the same principles, as it was in dynamics of a rarefied gas. Considered are the liquid particles in cells, and the process is divided into three main steps:

convective transfer

$$\left(v\frac{\partial f}{\partial x}\right);$$

turbulent dissipation of energy

$$\left(-\frac{1}{2\tau_1}\frac{\partial}{\partial v}(v'f)\right),$$

and the redistribution of energy

$$\left(\frac{f_M - f}{\tau_2}\right),$$

was numerically solved the problem of a dissipation of the spot, the energy of which was initially concentrated within the area with a characteristical radius r_0

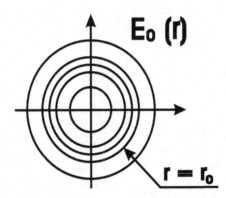

Fig. 10.10 Dissipation of the turbulent spot (initial area).

(see Fig. 10.10), a characteristical radius of the spot $r_*(t)$, and a density of turbulent energy $E_m(t)$ at the center of a spot. The initial data are the following:

$$E_0(r) = E_m^{(0)} \exp\left(-\frac{r^2}{r_0^2}\right),$$

$$f(0, r, v) = f_0(z, v),$$

$$f_0(r, v) = \left(\frac{3}{4\pi E_0}\right)^{3/2} \exp\left[-\frac{3v'^2}{4E_0(r)}\right].$$

Chapter 11

Studies of the Turbulent Flow of Fluid and Gas

11.1 Modeling of a Turbulent Transition within the Boundary Layer Using Monte Carlo Method (see Zharov, Tun Tun, Khlopkov[223])

On the basis of experimental data on the kinematics and dynamics of the turbulent spots (Emmons spots) at the flat plate in the incompressible fluid is proposed a statistical method of modeling of the flow within the transitional area of a boundary layer. This method provides the possibility to determine the intermittency, taking into account the effect of imposition of spots, one on another. This process permits to determine the forces at the surface of a plate, and the field of a flow in vicinity of a transitional zone, as soon as is known the field of a longitudinal component of the mean velocity within the developed turbulent boundary layer in its dependence on Reynolds number. Such an approach, in distinction from the multiparametrical models of a transition, permits to avoid the use of nonphysical values of parameters.

During the last time are made the serious attempts of the modeling of turbulent flows with the help of the methods of statistical modeling. In these attempts are used the models,[177] which are easily interpreted from the point of view of the theory of probability. By the consideration of the problem of the laminary-turbulent transition the role of such a model is played by kinematics and dynamics of the turbulent spots, which was proposed rather long time ago by Emmons.[212] The essence of that approach consists in the fact, that by certain critical Reynolds number is experimentally observed the appearance of point — like spots — embryos of a turbulence, which are growing as they move downstream, but keep unchanged their form, of which the parameters vary in accordance to the well-known simple laws. Within the spots the characteristics of flow are close to those of the developed turbulent flow by the Reynolds number, corresponding to the position of a spot. In accordance to that appears the possibility of a determination of the mean forces and fields within the traditional area. By the Reynolds number, close to the critical one, the spots appear in chaotical way both in space and in time. Using the information on the geometry of spots and on the statistics of their appearance, Emmons found the probability of such an event, that a certain point at the plate's surface is covered by the turbulent spot. This means that occurring is the so-called phenomenon of intermittency. By that it is assumed that the spots do not superimpose each other. Using the Monte Carlo method, the picture of that type might be modeled even when the superimposition is taken into account, but when known is the kinematics and dynamics of the single spots and the laws of their interaction. The data of that type might be found in a number of modern experimental studies,[205,213,215] of which the results are used in the present monograph.

For the closing of a model, it would be necessary to know the critical Reynolds number symbolizing the beginning of transition. There is an ample experimental material,[215] devoted to the connection between the parameters of oncoming flow and the Reynolds number of a transition, R_{1cr}. For that reason, at the initial stage of a modeling it would be expedient to use that information. Nevertheless, one should note the attempts to solve this problem not only phenomenologically (criterium e^n), but also to determine R_{1cr}, proceeding from the physics of development of the perturbations in the pretransitional area,[216] where this quantity is determined on the basis of a statistics of development of waves in the approximation of a three-wave resonance. Proposed in Refs. 217, 218 are the methods of determination of a dynamics of wave packets, which also might be used for the determination of R_{1cr}, and this we are going to do later. It is necessary to note, also, that there are numerous attempts to realize determination of the transitional area with the help of multiparametrical models of a turbulence. However, as it is proved by the practice of computations of that type, for the realistical determination of R_{1cr} it is necessary to preset the nonphysical values of the control parameters for such models.[219]

11.1.1 *Formulation of the probabilistical model of a transitional area*

Method of the determination of the density of distribution of probability of the flow's turbulization in the transitional area consists in the following. Beginning with R_{1cr} (x_0 — value of the longitudinal coordinate which corresponds to that Reynolds number) are formed the point-like spots with the frequency λ and in the correspondence to the density of distribution of probability in time, $\rho(t) = \lambda \exp(-\lambda t)$. These spots are formed in the statistically uniform way within the area $z \in [0, 1]$, $x \in [0, x^*]$. In the computations the value of x^* was taken equal to 0.1–0.2. By the downstream motion the spots chosen begin to grow (see Fig. 11.1).

The program of the development of spots in space and in time was created on the basis of the packet Mathematic.[220] Presented graphically in Fig. 11.2 is

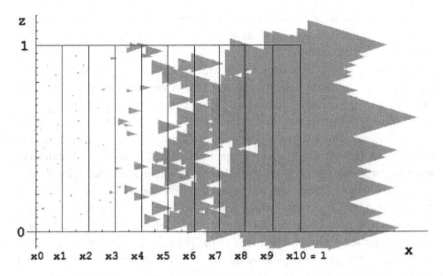

Fig. 11.1   Concrete realization of the kinematics of turbulent spots with $\lambda = 200$.

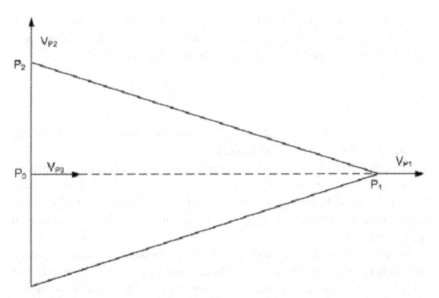

Fig. 11.2   Geometry of a turbulent spot and the speeds of propagation of its characteristical points P_0, P_2, P_3.

the scheme of evolution of the characteristical points (P_0, P_1, P_2) of the turbulent spot. The adopted form of the spots is the equisided triangle (this form is preserved during all the period of observation), the base of which moves with the velocity $V_{P_0} = 0.5U_\infty$, and the peak — with the velocity $V_{P_1} = 0.89U_\infty$, while the side peak — with the velocity $V_{P_2} = 0.1U_\infty$. Considered is the area between the line $x_0 = 0$ (see Fig. 11.1), where the spots are born, and the line $x_1 = 1$, where these spots are located so tightly, that the whole flow might be considered as turbulent one. This last line is determined by a numerical experiment. The area of flow $x \in [0, l]$ is divided into n equal parts (by the concrete computations was taken $n = 10$). Each one of the sub-areas is filled, accidentally, by the uniformly distributed over the sub-area points (in real computations 5000 points for the sub-area), and is determined the number $\frac{N_{spot}}{N_{full}} = f(x)$, where N_{spot} — the number of points, which are caught inside the spot, and N_{full} — the full number of points within the sub-area. The catch of the point by the spot, when the point (x, z) enters the area, occupied by the spot, is determined with the help of a condition

$$q(t, x, z, \mathbf{s}) = \frac{|-V_{P_0}t + x - s_x|}{(V_{P_1} - V_{P_0})t} + \frac{|z - s_z|}{V_{P_2}t} \leq 1 \bigcap x - V_{P_0}t - s_x \geq 0,$$

where $\mathbf{s} = (s_x, s_z)$ are the coordinates of a point P belonging to a spot. The sign "$\bigcap$" means the logical "and". Now, the condition for the point, meaning that it belongs, at least, to one spot, might be presented in the form

$$Q = \bigcup_{i=1}^{N} q(t, x, z, \mathbf{s}^i).$$

The sign "$\bigcup$" means the logical "or". The values of f obtained as a result, are afterwards averaged over several realizations.

Presented in Fig. 11.3 is the singling out of the part of those spots, which are found within the strip chosen. It should be noted that here, in distinction of the work by Emmons, the density of distribution of the probability of turbulization of the flow (over the relative area S_{spot}/S_{full}) is determined with taking into account the over covering of the spots.

Presented in Fig. 11.4 is the density f of a distribution of probability of the turbulization of a flow within the transitional area, by $\lambda = 200$ (points). Drawn in the same graph is the approximational curve $f_1 = 0.5(1 + \mathrm{erf}((x - a)/b))$, of which the parameters are found by the method of least squares, and are well approximating the numerical results. For comparison, the same method was used for determination of a constant in the expression $f = 1 - \exp[-x^2/x_0^2]$; that constant was obtained in Ref. 212.

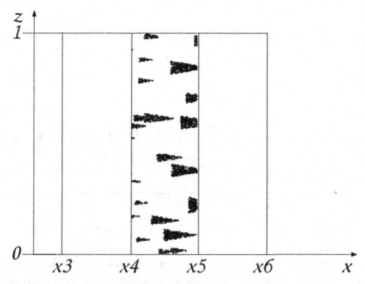

Fig. 11.3 Division of the computational area into sub-areas and the determination of area occupied by the spots within the sub-area picked out by the Monte Carlo method.

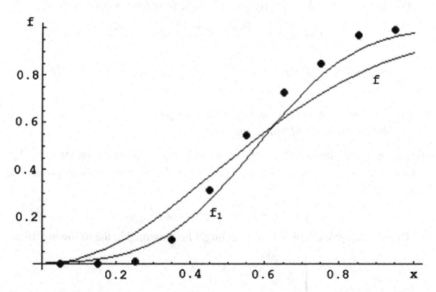

Fig. 11.4 Comparison of the results of numerical analysis of the turbulization degree within the transitional area according to Emmons ($f = 1 - \exp[-x^2/x_0^2]$) and with the help of approximating function $f_1 = 0.5(1 - \mathrm{erf}((x - a)/b))$.

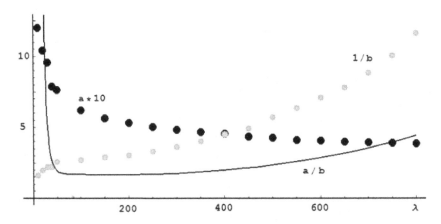

Fig. 11.5   Dependence of the coefficients a, $1/b$, a/b on the speed of birth of the turbulent spots λ.

As it is seen, the approximational function displays the worse description both of the Emmons approximation and of the numerical data. From the comparison of the numerical results with an approximational function are determined the quantities a and b, as the functions of λ. The results are shown in Fig. 11.5.

The curves obtained are approximated by the following polynomials:

$$a = 0.313976 + \frac{127587}{\lambda^3} - \frac{4607.78}{\lambda^2} + \frac{63.5291}{\lambda},$$

$$\frac{1}{b} = 2.39998 + 0.00286859\lambda + 7.1233610^{-7}\lambda^2 + 1.230493910^{-8}\lambda^3.$$

11.1.2   *Determination of the coefficient of drag for a plate by the presence of transition*

The curve of variation of the local coefficient of drag should coincide with the experimental one.[215] The local coefficient of the plate's drag might be presented in the form

$$c'_f = (1 - f(R))c'_{f\text{lam}} + f(R)c'_{f\text{turb}}.$$

Then the integral coefficient of drag might be presented in the following form:

$$c_f = \frac{1}{R_l}\int_0^{R_l} c'_f dR_l = \begin{cases} \dfrac{1}{R_l}\displaystyle\int_0^{R_l} c'_{f\text{lam}} dR_l, & R_l < r_k, \\[3mm] \dfrac{1}{R_l}\displaystyle\int_0^{r_k} c'_{f\text{lam}} dR_l + \dfrac{1}{R_l}\displaystyle\int_0^{r_k} ((1-f)c'_{f\text{lam}} + fc'_{f\text{turb}})dR_l, \\[3mm] \hfill R_l \geq r_k, \end{cases}$$

$$= \begin{cases} \dfrac{1.328}{\sqrt{R_l}}, & R_l < r_k, \\[3mm] \dfrac{r_k}{R_l}\dfrac{1.328}{\sqrt{R_l}} + \dfrac{1}{R_l}\displaystyle\int_{r_k}^{R_l}\left((1-f)\dfrac{0.664}{\sqrt{R}} + f\dfrac{0.0576}{\sqrt{R}}\right)dR, \\[3mm] & R_l \geq r_k. \end{cases}$$

The expression for f, when presented in the form, convenient for these purposes, might be written down as

$$f = 0.5\left(1 + \mathrm{erf}\left[\frac{a}{r_0}*(R - r_k - b*r_0)\right]\right).$$

Here R is the local Reynolds number of the point at the plate, r_0, r_k — the auxiliary quantities. Making comparison with the graph from the Ref. 215, one is able to determine the quantities a, b, r_0, r_k. Algorithm for the determination of these quantities is the following. First of all, one determines the dependence of a ratio a/b on λ. This dependence is presented in graphical form in the Fig. 11.5.

It is to be noted that this dependence has a minimum at the point $\lambda = 132$. Let us determine values of a and $1/b$ at that point according to their approximations over λ. One obtains

$$a = 0.55; \quad b = 2.73.$$

The quantity r_k is determined through the number $R_{1\mathrm{cr}}$, which corresponds to a beginning of transition. After that, the quantity r_0 is chosen with the help of a condition of the best coincidence with the experimental data:

$$r_0 = 10^4; \quad r_k = 4 \times 10^5.$$

As a result, one obtains the dependence (see Fig. 11.6) of the coefficient of a plate's drag on the Reynolds number, which confirms the good coincidence with experimental data of Ref. 215.

11.1.3 Conclusions

As it is seen from the preceding material, the methods of statistical modeling permit to determine kinematics of the turbulent spots within the transitional area of a flow about the plate, at the different moments of time. In the essence of matter, the complicated area, occupied by the turbulent spots, is modeling the bearer of a solution of the full dynamical problem for the Navier–Stokes equations. The kinematics of spots is determined by the frequency of their birthers, λ. As soon as this quantity is known, it would be possible to determine a local degree of the flow's turbulization, with the help of which one might calculate the various characteristics of a turbulent flow within the transitional area, for example: to determine local and integral forces, or to build up the mean field of flow within that area. It is to be noted that in distinction of the study by Emmons, this quantity is determined taking into account the effect of the

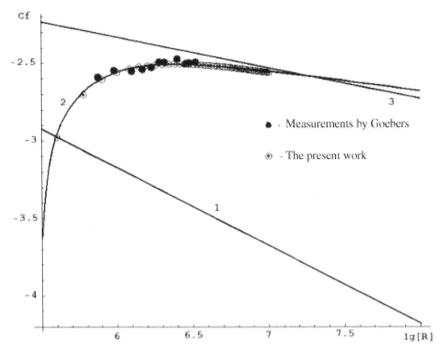

Fig. 11.6   Comparison of the total drag of a plate in dependence on the Reynolds number of plate, obtained numerically in accordance to data.[215]

overcovering of spots. The presence of a minimum in the dependence of $a(\lambda)/b(\lambda)$ on λ means, most probably, that by the lesser values of λ that mechanism of a transition, which is connected with the development of turbulent spots, is not already suitable.

11.2   Study of the Dissipation of Turbulent Spots (see Belotserkovskii, Yanitskii, Bukin[12,221])

The problem was formulated in the following way. Considered is the embryo of a turbulence in the form of a certain spot (see Figs. 11.7–11.12) with the proper initial and boundary conditions. After that is considered the temporal evolution of the same spot in accordance to the "fluidical" model of turbulence (see Sec. 10.4 of Chap. 10).

The kinetic models of a turbulence are more informative, because they describe the pulsation at the level of a distribution function. Such an approach to the description of a turbulence seems to be a prospective one, since it permits to consider the large-scale turbulent processes directly with the help of the equations of transfer, while the small-scale pulsations are studied through the direct statistical modeling.

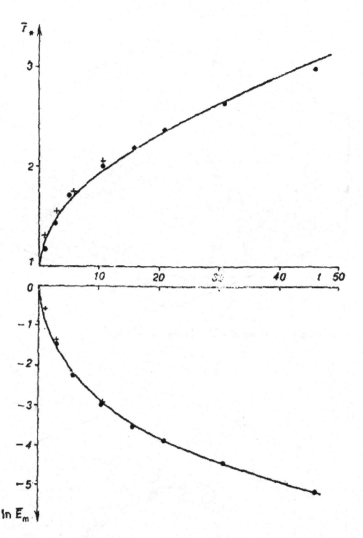

Fig. 11.7 Dissipation of a turbulent spot — data of experiment (Naudascher), + direct modeling.

11.3 Evolution of the Vortical System in the Rarefied Gas (see Rovenskaya, Voronich, Zharov[222])

Considered is the Boltzmann equation:

$$\frac{\partial f}{\partial t} + \xi_x \frac{\partial f}{\partial x} + \xi_y \frac{\partial f}{\partial y} + \xi_z \frac{\partial f}{\partial z} = J(f, f) = \int (f' f_1' - f f_1) g b \, db \, d\varepsilon \, d\vec{\xi}_1,$$

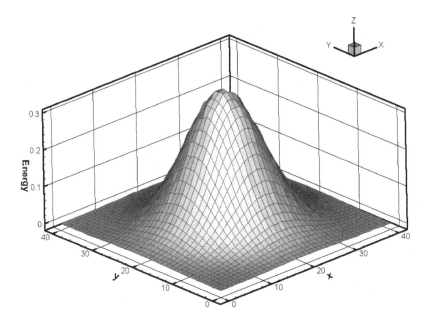

Fig. 11.8   Distribution of a specific energy in turbulent spot, $t = 0$.

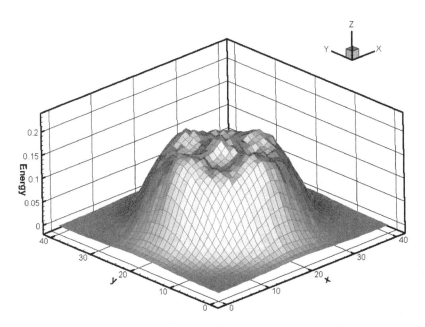

Fig. 11.9   Distribution of a specific energy, $t = 40$.

Distribution of energies, *t* = 0

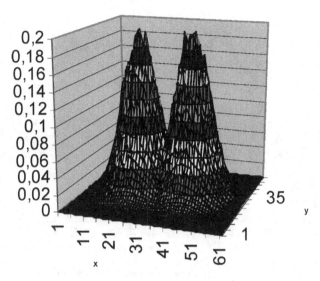

Fig. 11.10   Functions of distribution of a turbulent energy of the interacting spots.

t = 40

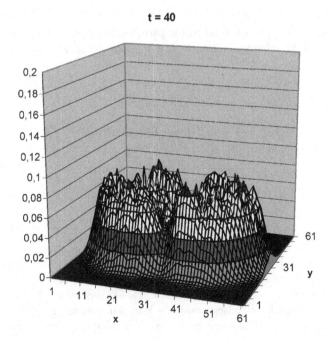

Fig. 11.11   Functions of distribution of a turbulent energy of the interacting spots.

Cross-section for y = 31

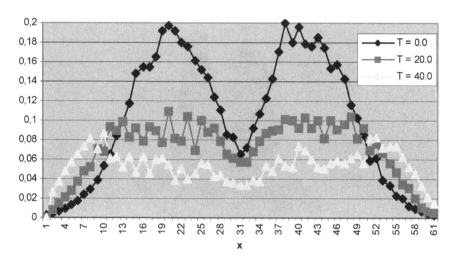

Fig. 11.12   Interaction of two turbulent spots (distribution of energies).

where $f = f(t, \vec{x}, \vec{\xi})$, $f_1 = f(t, \vec{x}, \vec{\xi}_1)$, $f' = f(t, \vec{x}, \vec{\xi}')$, $f_1' = f(t, \vec{x}, \vec{\xi}_1')$ while the velocities $\vec{\xi}$, $\vec{\xi}_1$, $\vec{\xi}'$, $\vec{\xi}_1'$ correspond to the particles before collision and after that, $\vec{g} = \vec{\xi}_1 - \vec{\xi}$ is the relative velocity, b — the aiming distance, ε — the azimuthal angle.

Method of splitting

$$\frac{\partial f}{\partial t} = J(f, f) \text{ — the relaxation,}$$

$$\frac{\partial f}{\partial t} + \vec{\xi}\frac{\partial f}{\partial \vec{x}} = 0 \text{ — the flying over.}$$

Initial conditions

$$f_0(t, \vec{x}, \vec{\xi}) = n(t, \vec{x}) \left(\frac{m}{2\pi k_B T(t, \vec{x})}\right)^{3/2} \exp\left(-\frac{m(\vec{\xi} - \vec{V}(t, \vec{x}))^2}{2k_B T(t, \vec{x})}\right),$$

where n — concentration of the particles, T — temperature, $\vec{\xi}$ — velocity vector of the particle, m — mass of the particle, k_B — constant by Boltzmann. Here is used the method by Belotserkovskii — Yanitskii, adapted for the solution of unsteady problems and applying the model of solid spheres to the modeling of collisions.

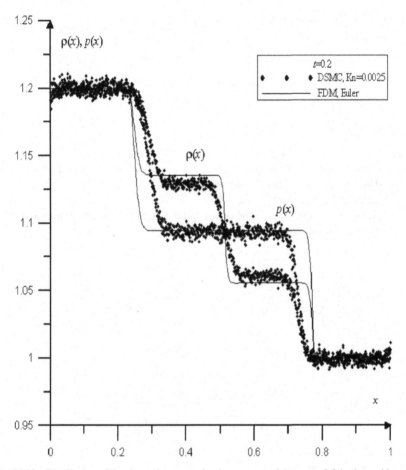

Fig. 11.13   Distribution of density and pressure for the moment of time $t = 0.2$ in the problem of rupture's breakdown, by $Kn = 0.0025$.

Check of the algorithm was carried out in the respect to the one-dimensional problem of the collapse of a breakup:

$$(\rho_L, u_L, T_L) = (1.2, 0, 1),$$

$$(\rho_R, u_R, T_R) = (1, 0, 1).$$

At the boundaries is realized the specular reflection of the particles. Regims: $Kn \to 0$, $Kn \to \infty$, $Kn = 0.01$, $Kn = 0.005$, $Kn = 0.0025$.

The comparison was made both with the theoretical results and with the computations on the basis of finite-difference methods (Fig. 11.13).

Formulation of the problem by Taylor-Green
Initial and boundary conditions:

$$u_0(x, y) = A \sin(2\pi x) \cos(2\pi y), \quad v_0(x, y) = B \cos(2\pi x) \sin(2\pi y),$$

$$\rho_0(x, y) = 1 + C \sin(2\pi x) \sin(2\pi y), \quad T_0(x, y) = 1 + D \cos(2\pi x) \cos(2\pi y),$$

$$0 < x < 1, 0 < y < 1, A = 0.5, B = -0.4,$$

$$C = D = 0.1, F(x + 1, y + 1) = F(x, y).$$

Mmax $= 0.39$ — the Mach number, as determined the initial field, $\gamma = 5/3$ — the adiabatic factor in the monatomic gas, $Kn = 0.01$ — the Knudsen number, $Re = \text{Mmax}/Kn = 39$ — the Reynolds number.

The network used is $110 \times 110 \times 1$.

At the boundaries is realized the periodical transfer of particles, similar to that in the problem of evolution of the vortical system (Figs. 11.14–11.20).

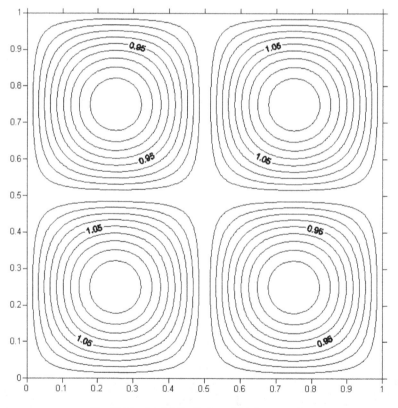

Fig. 11.14   Spatial distribution of density at $t = 0$.

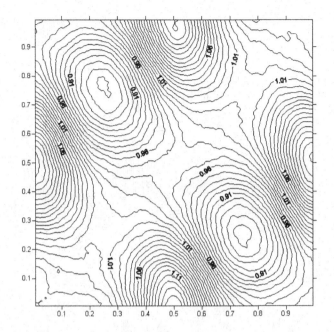

Fig. 11.15   Spatial distribution of density at $t = 0.1$.

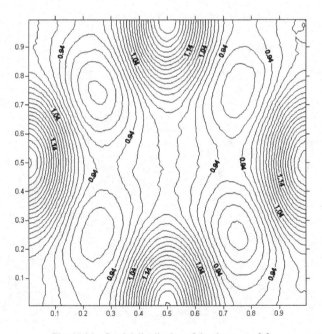

Fig. 11.16   Spatial distribution of density at $t = 0.2$.

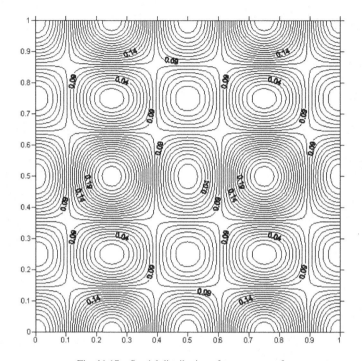

Fig. 11.17   Spatial distribution of energy at $t = 0$.

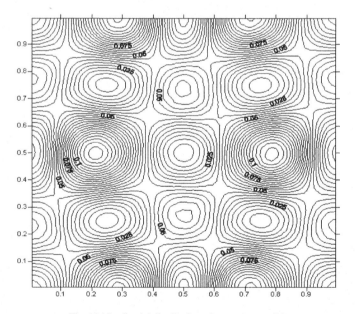

Fig. 11.18   Spatial distribution of energy at $t = 0.1$.

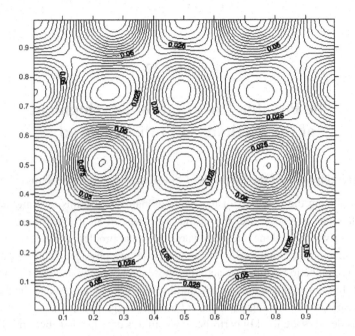

Fig. 11.19   Spatial distribution of energy at $t = 0.2$.

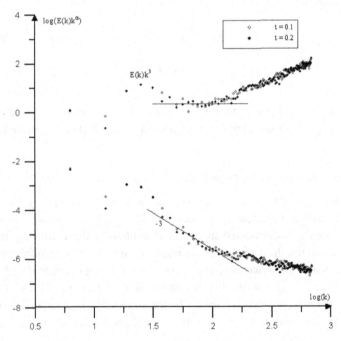

Fig. 11.20   Spectrum of energy at $t = 0.1$ and $t = 0.2$ in the problem of evolution of the vortical system.

Chapter 12

The Possible Directions of Development of the Methods of Statistical Study

12.1 Development of the Methods of Solution of Linear Problems

When analyzing the materials of the International Symposiums on Rarefied Gas Dynamics, one is able to see that in the last times, in all the world around has been strongly increased the interest towards the linear problems in various areas of physics, electronics connected with nanotechnologies, physical chemistry, and so on. In this connection acquire the actuality those methods, which were presented in Chap. 2. In particular, the method of singling out of the main part and the method of solution of the linear, integro-differential kinetic equation.

Considered is the weakly perturbed flow. In this case the distribution function proves to be little different from some characteristical Maxwellian distribution function, that is

$$f = f_{00}(1 + \phi), \tag{12.1.1}$$

$$f_{00} = n_0 \left(\frac{m}{2\pi k T_0} \right)^{3/2} \exp \left(\frac{m}{2k T_0} \xi^2 \right), \quad \varphi \ll 1,$$

where m is the mass of a molecule, k — constant by Boltzmann, ξ — velocity of a molecule, φ — a small addition, the square of which might be neglected.

12.1.1 *Singling out of the main part*

The process of transfer might be considered as an uniform chain by Markov, of which the links are representing the locations of particles immediately before the collision. Since here considered are the small additions to the distribution function and to its moments, the statistical errors might occur to be of the same orders as the quantities, which are looked for. Therefore, the wandering of the trial molecule is gambled over the equilibrium distribution function, and subtracted from the total transfer is that one, which is carried out by the absolutely Maxwellian distribution. In this case, the mean transfer of a certain molecular quality through any chosen plane is equal to

$$\Psi = \frac{1}{N} \sum_{\alpha=1}^{N} \left(\frac{W}{W_0} - 1 \right) \sum_{\beta(\alpha)} \psi_{\alpha\beta}, \qquad (12.1.2)$$

where N is the number of trajectories, W and W_0 — the probabilities of flying of the particle along the trajectory α, for the perturbed flow and for unperturbed one, correspondingly. The quantity $\beta(\alpha)$ is the number of intersections of the present plane by the trajectory.

The wandering of a particle is realized in following ways.

1. Gambled at one of the boundaries are the velocities of the molecule, which is flying in, with the density of probability proportional to

$$f_{00}(\vec{\xi}) \cdot (\vec{\xi}, \vec{n}).$$

2. The collisionless flight during the time interval τ and collision within the element $d\tau$ is gambled with probability

$$d\tau \exp\left(- \int \left[\int f_{00} g \sigma d\vec{\xi}_1 \right] d\tau \right) \int f_{00} g \sigma d\vec{\xi}_1.$$

3. The velocity of a molecule-partner is gambled with the density of probability $g\sigma f_{00}(\vec{\xi}_1)$.

4. The collision is gambled in accordance to the law of interaction of the molecules considered, and so on, beginning with the item 2 and until the moment, when the molecule leaves the area of a flow.

The ratio W/W_0 is found in accordance to the formula

$$\frac{W}{W_0} = \frac{F_w}{F_{w0}} \frac{P}{P_0} \frac{F}{F_0} \frac{Q}{Q_0} \cdots, \qquad (12.1.3)$$

where F_w and F_{w0} are the probabilities for the molecule of the flying out of the boundary of the? correspondingly, perturbed flow and unperturbed one, P and P_0 — probabilities of the free flight during time τ, F and F_0 — probabilities of the collision within the interval $d\tau$ with a molecule flying with a velocity $\vec{\xi}_1$, Q and Q_0 — probabilities of the collision with a prescribed aiming distance. As soon as one uses the linearity of a problem, expression (12.1.3) might be presented in the form of a sum

1.

$$F_w = \frac{(\vec{\xi}_1, n) f_0 (1 + \varphi_w) d\vec{\xi}}{N_u}, \quad F_{w0} = \frac{(\vec{\xi}_1, \vec{n}) f_0 d\vec{\xi}}{N_0},$$

$$N_u = \int (\vec{\xi}_1, \vec{n}) f d\vec{\xi}, \quad N_0 = \int (\vec{\xi}_1, \vec{n}) f_0 d\vec{\xi}.$$

2.

$$P = \exp\left(-\iint f_1 g\sigma d\vec{\xi}_1 d\tau\right) d\tau \int f_1 g\sigma d\vec{\xi}_1$$

$$= \exp\left(-\iint f_{01}(1+\varphi_1)g\sigma d\vec{\xi}_1 d\tau\right) d\tau \int f_{01}(1+\varphi_1)g\sigma d\xi_1.$$

Let us introduce the notations: $K_0 = \int f_0 g\sigma d\vec{\xi}_1$, $K_0 k = \int f_0 \varphi g\sigma d\vec{\xi}_1$.
Then, after using the condition $k \ll 1$, one obtains

$$P = \exp\left(-K_0\tau - K_0 \int k d\tau\right) d\tau(K_0 + K_0 k)$$

$$= \exp(-K_0\tau)\left(1 - K_0 \int k d\tau\right) d\tau K_0(1+k).$$

Similarly, for $P_0 = \exp(-K_0\tau) d\tau K_0$.

3.

$$F = \frac{f_1 g\sigma\xi_0}{\int f_1 g\sigma d\vec{\xi}_1}, \quad \frac{f_{01}(1+\varphi_1)g\sigma d\vec{\xi}_1}{FK_0(1+k)},$$

$$F_0 = \frac{f_{01} g\sigma\xi_1}{K_0}.$$

4. Quantities Q and Q_0 are equal for the equilibrium trajectory and actual one.

After substitution of the expressions for the probabilities obtained into expression (12.1.3) one will find that

$$\frac{W}{W_0} = \frac{N_0}{N_u}(1+\varphi_w)\left(1 - K_0 \int k d\tau\right)(1+\varphi_1)\cdots$$

$$= \frac{N_0}{N}\left(1 + \varphi_w - K_0 \int k d\tau + \varphi_1 + \cdots\right). \tag{12.1.4}$$

Taking into account that from the surface of a boundary flies in the totality of SN_u molecules, one will obtain, ultimately,

$$\Psi = \frac{SN_0}{N} \sum_{\alpha=1}^{N}\left(\varphi_w - K_0 \int k d\tau + \varphi_1 + \cdots\right)\sum_{\beta(\alpha)} \psi\alpha\beta, \tag{12.1.5}$$

where S is the area of a boundary.

From the above considerations it follows that for the gambling of the trial molecule it is necessary to know the distribution function for the field particles. Therefore, the problem is solved by the method of successive approximations. For the initial approximation it would be possible to take any function, for example — zero. The process of computation goes on till the moment, when $\varphi^{(n)} \approx \varphi^{(n-1)}$ with a prescribed degree of accuracy.

12.1.2 *Solution of the linearized kinetic equation*

Considered is the linearized Boltzmann equation:

$$\xi_x \frac{\partial \varphi}{\partial x} + \xi_y \frac{\partial \varphi}{\partial y} + \xi_z \frac{\partial \varphi}{\partial z} = -\varphi k(v) + \int K(\xi, \xi_1) \varphi_1 d\xi_1, \tag{12.1.6}$$

where $\xi = \xi \sqrt{\frac{m}{2kT}}$.

The analytical expressions for the molecules, interacting in accordance to the law of solid spheres, have the form

$$k = a \int g e^{-v_1^2} dv_1, \tag{12.1.7}$$

$$K = B e^{-v^2} \left(g - \frac{2}{g} e^{D^2} \right), \tag{12.1.8}$$

where $g = v_1 - v$, a and B are the constant coefficients.

Carrying out the integration of Eq. (12.1.6) along the trajectories, one is coming to the linear integral equation

$$\varphi = \varphi_r + \int K \varphi_1 dv_1 dl, \tag{12.1.9}$$

where

$$\varphi_r = \varphi_w e^{-\frac{k}{v}(l-l_0)}, \tag{12.1.10}$$

$$K = \frac{1}{v} K e^{-\frac{k}{v}(l_1-l_0)}. \tag{12.1.11}$$

Equation (12.1.9) might be solved with the help of statistical procedure by Uhlam–Neumann. In the general case, the computation by Monte Carlo method is reduced to a calculation of integrals. The role of such integrals is played by the mathematical expectations of accidental quantities, which appear as estimates of something; in other words, carried out is the estimate of the integral of the type of Lebesgue–Stillties over some probable measure:

$$l = \int \psi(x) u dx, \tag{12.1.12}$$

with the help of a mean arithmetical quantity over the number of tests,

$$\frac{1}{N} \sum_{i=1}^{N} \psi(x_i). \tag{12.1.13}$$

Turning to Eq. (2.3.5), it is to be noted that this equation is connected with the uniform chain by Markov, prescribed by the density of initial distribution and by a transitional density $p(x \to y)$. The selectional trajectory of that chain is built up in accordance to the initial density of probability and to the transitional one. Then the estimate of the functional, which is looked for, will appear in the form

$$W/W_0,$$

where

$$\frac{W}{W_0} = \frac{N}{N_0}\left(1 + \varphi_w - An_0\int v d\tau + v + \varphi_w + \cdots\right), \quad \varphi_w,$$

where k is the number of collisions among the trajectory.

Thus, when the analytical expression for the kernel is known, one will be able to build up the simple algorithm for calculation of the various functional for a solution of the integral equation.

12.2 Use of the Possibilities of the Model Equations

12.2.1 *Decrease of the volume of operative memory*

Presented in Chaps. 1 and 3 were the possible directions of development of the methods of statistical modeling, using the specific features of model equations.

Used as the basic equations were those presented in Chap. 1:

— Model equation by Krook:

$$\frac{df}{dt} = v(f_0 - f), \quad f_0 = n\left(\frac{m}{2\pi kT}\right)^{3/2} e^{-\frac{m}{2kT}(\bar{\xi}-\bar{v})^2}, \tag{12.2.1}$$

where v is the frequency of collisions, f_0 — the equilibrium distribution function.
— Ellipsoidal model by Holway[19]:

$$\frac{df}{dt} = v(f_e - f), \tag{12.2.2}$$

f_e is the ellipsoidal distribution function.
— Approximational model by Shakhov[20]:

$$\frac{df}{dt} = v(f^+ - f), \quad f^+ = f_0\left[1 + \frac{4}{5}(1 - Pr)s_\alpha c_\alpha\left(c^2 - \frac{5}{2}\right)\right],$$
$$s_i = \frac{1}{n}\int c_i c^2 f d\bar{\xi}, \tag{12.2.3}$$

where c is the dimensionless molecular thermal velocity and is the Prandtl number.

The principial feature of the model Eq. (12.2.1), in distinction of the Boltzmann equation, consists of such a notion, that this equation assumes the distribution function of molecules after the collision to be the most probable by the prescribed numbers of colliding particles, their momenta and their energies. The knowledge of macroparameters is completely determinating either the state of a system or the trajectory of trial particle, since the velocities of molecules after the collision are distributed with a function

$$f_0 = n \left(\frac{m}{2\pi kT} \right)^{3/2} e^{-\frac{m}{2kT}(\xi - u)^2}.$$

Thus, appears the possibility to construct the trajectory of trial particle within a physical space, divided into such cells, within which at each of the iterations is kept the information not on the distribution function, but just only on the macroparameters.

In principle, the reasoning of the same type might be conducted for Eq. (1.1.4), too, and for that equation the molecular distribution function after the collision may be assumed to be equal to

$$f^+ = f_0 \left[1 + \frac{4}{5}(1 - Pr)s_\alpha c_\alpha \left(c^2 - \frac{5}{2} \right) \right],$$

and in order to come to such a situation it would be necessary, in the general case, to memorize within each cell not only density, velocity, and temperature, but also the stress tensor and the heat flow.

12.2.2   Unsteady statistical modeling for the solution of model kinetic equations

In Chap. 3 were considered the possibilities of a scheme of direct statistical modeling on the basis of an approximate kinetic equation by Krook:

$$\begin{cases} f^{n+1/2} = f^n + J\dfrac{\Delta t}{2}, \\ f^{n+1} = f^n + J^{n+1/2}\Delta t. \end{cases} \tag{12.2.4}$$

From (3.3.3) one obtains

$$f^{n+1/2} = f^n \left(1 - v^n \frac{\Delta t}{2} \right) + v^n \frac{\Delta t}{2} f_0^n, \tag{12.2.5}$$

$$\begin{aligned} f^{n+1} &= f^n + (f_0^{n+1/2} - f^{n-1/2})v^{n+1/2}\Delta t \\ &= f^n - f^{n+1/2}v^{n+1/2}\Delta t + f_0^{n+1/2}v^{n+1/2}\Delta t. \end{aligned} \tag{12.2.6}$$

From (12.2.5) one obtains

$$f^n = \frac{f^{n+1/2} - v^n \frac{\Delta t}{2} f_0^n}{1 - v^n \frac{\Delta t}{2}},$$

and after substitution into (12.2.5) one obtains

$$f^{n+1} = \frac{f^{n+1/2} - v^n \frac{\Delta t}{2} f_0^n}{1 - v^n \frac{\Delta t}{2}} - f^{n+1/2} v^{n+1/2} \Delta t + f_0^{n+1/2} v^{n+1/2} \Delta t$$

$$= f^{n+1/2} - v^n \frac{\Delta t}{2} f_0 - f^{n+1/2} v^{n+1/2} \Delta t \left(1 - v^n \frac{\Delta t}{2}\right)$$

$$+ \frac{f_0^{n+1/2} v^{n+1/2} \Delta t \left(1 - v^n \frac{\Delta t}{2}\right)}{1 - v^n \frac{\Delta t}{2}}.$$

Neglecting the terms of the smallness $0(\Delta t^2)$ and expanding in series

$$\frac{1}{1 - v^n \frac{\Delta t}{2}} = \left(1 - v^n \frac{\Delta t}{2}\right)^{-1} = 1 - v^n \frac{\Delta t}{2} + 0(\Delta t^2),$$

one obtains

$$f^{n+1} = f^{n+1/2}[1 - (v^{n+1/2} - v^n/2)\Delta t] + (v^{n+1/2} - v^n/2) f_0^{n+1/2} \Delta t. \quad (12.2.7)$$

Thus, the process as a whole might be modeled in such a way. Let be conducted the flight-over, within the time $\frac{\Delta t^*}{2}$, let be memorized the frequency of collisions $v_j^{n+1/2}$ and let the particles be collide in accordance to a scheme described above. Once again, let us choose the accidental number R_i (distributed uniformly within the interval between 0 and 1), and let us change the velocity of ith particle, if $(v_j^{n+1/2} - v^n/2)\Delta t \geq R_i$, or let us not change it, if $(v_j^{n+1/i} - v_j^{n+1}/2)\Delta t \prec R_i$, where v_j^{n+1} is the frequency of collisions in jth cell, at $(n+1)$th step in time Δt^* (Fig. 12.1).

The iterational scheme, presented in such a form, is, practically, excluding the splitting in time Δt^* for two events: transfer of the particles and their collision. Proceedings in that way, one has no limitation for Δt^* in connection with the criterium of stability. Here the quantity Δt^* was chosen in the form

$$\Delta t^* = \frac{\alpha Kn}{\sqrt{3RT^*}}, \quad (12.2.8)$$

the coefficient α being varied from 0.5 till 1. Comparison of the two versions of computations for the problem of heat transfer, using the older scheme and the newer one, shows, that the process, based on a new scheme, converges approximately on one order more rapidly (see Fig. 12.2).

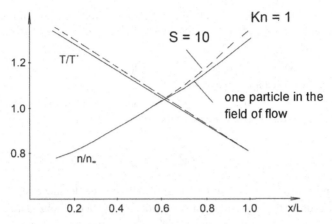

Fig. 12.1   Distribution of a density n and a temperature T for the cases of 10 particles in cell and 1 particle.

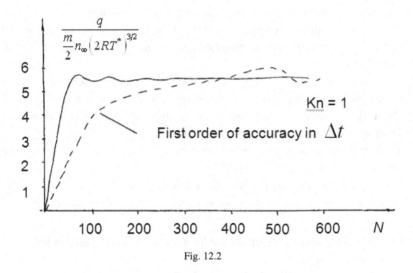

Fig. 12.2

12.3   Modeling of the Flows of Continuous Medium

The study of problems of the rarefied gas dynamics is connected, as a rule, with a solution of the kinetic equations for the distribution function. The wide application of the Monte Carlo methods in this area of science is stipulated by the complicated, multi-dimensional structure of the kinetic equations, by the probabilistical nature of the processes studied, and by the abundance of information contained in the distribution function. The statistical description of the fluid and gas, which was presented in the Introduction, permits to extend the Monte Carlo approach into the area of continuous medium, as well, were such an approach to a solution appears to be a nontraditional one, but, nevertheless, permits to use the advantages of the methods of statistical modeling by the study of the ideal flows (Sec. 4.3 of Chap. 4), of the viscous flows (Sec. 4.1 of Chap. 4 and Sec. 5.6 of Chap. 5) and the turbulent flows (Chaps. 10 and 11).

Let us remind the main properties of Monte Carlo methods. Alongside with the "heavy" dependence of error on the number of trials,

$$I = \frac{1}{N} \sum_{i=1}^{N} \xi_i \pm \frac{3\sigma}{\sqrt{N}}, \qquad (12.3.1)$$

with which, as it was already shown in this book, one could successfully fight in general case, using the physical nature of a process modeled, Monte Carlo methods possess a great variety of the remarkable peculiarities which are not proper to the regular methods.

(1) the error does not react upon the dimensionality of a problem;
(2) the simple structure of a computational algorithm (N times repeated, belonging to one type calculations of the realizations of accidental quantity of accidental trajectory);
(3) the construction of an accidental quantity ξ might be, in general case, based on the physical nature of a process, and will not demand the obligatory, as in the regular methods, formulation of the equation, which property becomes to be more and more actual for the study of modern problems;
(4) the united statistical procedure permits to build up the general approach to the modeling of various regimes of a flow;
(5) the single-typed structure of a computational algorithm permits to find a simple way to the construction of effective schemes of the parallel computations.

12.3.1 *Modeling of the continuous medium on the basis of a molecular distribution function*

One of the main ways of a statistical modeling is reduced to the splitting of the process into two, quite evident, physical stages.

— *molecular transfer*

$$\frac{\delta f}{\delta t} = -\xi \nabla f, \qquad (12.3.2)$$

— *relaxation*

$$\frac{\delta f}{\delta t} = J(f). \qquad (12.3.3)$$

For various classes of the flow of continuous medium the distribution function is, as a rule, known. For the ideal fluid it is a locally equilibrium distribution function:

$$f_0 = n \left(\frac{m}{2\pi kT} \right)^{3/2} e^{-\frac{m}{2kT}(\bar{\xi}-\bar{v})^2}.$$

For the viscous fluid it is the approximation by Navier–Stokes, obtained with the help of expansion of the distribution function in terms of a small parameter, which expansion, besides of density, velocity and temperature, contains the stress tensor P_{ij} and the heat flow q_i:

$$f_{NS} = f_0[1 + a(\xi)P_{ij} + b(\xi)q_i].$$

And in the most general case the distribution function might be presented in the form

$$f = F(\xi, M^{(0)}, \ldots, M^{(k)}), \tag{12.3.4}$$

where $M^{(k)}$ is the kinetical moment of the kth degree.

Thus, the first stage might be modeled by a molecular transfer, whereas at the stage of relaxation (12.3.3) one could pass to the macro level of a description, and due to such a way of action is essentially simplified the most complicated part of a solution of kinetic problems — that of the calculation of the integral of collisions:

$$\frac{\delta M^{(k)}}{\delta t} = \alpha M^{(k)}, \quad M^{(j+1)} = M^{(j+1/2)}(1 + \alpha \Delta t), \tag{12.3.5}$$

where M^j and $M^{j+1/2}$ — the moments after the stage of relaxation and of the transfer, correspondingly, the coefficient $\alpha = 0$ for the equations, containing the concentration n, the mean velocity u, and the temperature T within the cell; $\alpha = -1/\tau_P$ for the elements of a stress tensor P_{ij}; $\alpha = -2/(3\tau_P)$ for the components of a heat flow q_i, where τ_P — relaxation time.

Such an approach has recommended itself in sufficiently effective way for the modeling of the flows of an ideal gas, i.e. for the solution of Euler equations (Sec. 4.3 of Chaps. 4 and 9). It would be natural to assume that the extension of the algorithm of modeling of the stage of collisions on the basis of distribution function f_0 towards the distribution function f_{NS} will permit to pass to the solution of Navier–Stokes equations.

12.3.2 Modeling of the flows of continuous medium with the help of "fluids" (liquid particles)

Considered is a statistical procedure of the modeling of gaseous flows by some set of particles, totality of which determines the state of a medium (Sec. 4.1 of Chap. 4).

The field of a flow is divided into cells, the particles are located in the field and are endowed by a set of indices, as, for example, by the mass m, by the velocity ξ, by the coordinate x. For the rarefied gas a totality of such indices is sufficient for the realization, by the proper number of particles N within the cell, of the characterization of a distribution function, and, after the averaging over the particles, to determine the macroparameters. With the areas of flow, described by the differential equations for the moments of a distribution function, those particles, which determine the state of a gas, possess somewhat different set of the indices ψ. In this case, the velocities of particles in cell correspond to the velocity of a flow, and, moreover, the particles possess the internal energy ε. By analogy with a kinetic distribution function one might consider the distribution function of "fluids" in the form

$$f \approx \prod_{i=1}^{N} \delta(\xi_i - u)\delta(\varepsilon_i - E),$$

where u and E are the velocity and the internal energy of gas within the cell. Then the determination of macroparameters is realized by a standard way:

$$\Psi = (f, \psi) \approx \frac{1}{N} \sum \psi_i.$$

In those gaseous flows, which are described by the moment's equations of the higher order, the particles might possess the larger set of internal indices $P_{ij}, q_i, \ldots$.

The trajectories of particles are, generally speaking, described by the differential equations of motion

$$m\frac{d\vec{\xi}_i}{dt} = \sum_j F_{ij}, \quad \frac{d\vec{x}_i}{dt} = \vec{\xi}_i, \tag{12.3.6}$$

where F_{ij} is a force acting on the particle j from the part of the particles i. The evolution of a history of the particles in time occurs within the succession of the finite intervals of a duration Δt, and during that period are changing the velocities and internal indices of the particles, while afterwards is realized the transfer of these indices along the trajectories.

In the case of equations of moments the term, defining the force, in (12.3.6) is determined by the gradients of macroparameters. The trajectories of particles are constructed in accordance to a scheme:

$$\xi_j^{n+1} = \xi_j^{n1} + \phi(\vec{\xi})\Delta t,$$

$$\vec{x}_j^{n+1} = x_j^n + \left(\xi_j^n + \xi_j^{n+1}\right)\frac{\Delta t}{2}, \tag{12.3.7}$$

while the variation of internal indices ε is also stipulated by the gradient terms in the corresponding equations:

$$\varepsilon_j^{n+1} = \varepsilon_j^n + \phi^n(\vec{\xi})\Delta t.$$

12.3.3 Method of solution of the Navier–Stokes equations

The statistical method of the modeling of the viscous, incompressible fluid with the help of a system of discrete vortices, proposed in the Chap. 6 for the Navier–Stokes equations in the modified form by Helmholtz, seems to be rather prospective one:

$$\frac{\partial \Omega}{\partial t} + V_x \frac{\partial \Omega}{\partial x} + V_y \frac{\partial \Omega}{\partial y} = \nu \Delta \Omega,$$

$$\Omega = \operatorname{rot}\vec{V}; \quad \operatorname{div}\vec{V} = 0.$$

(12.3.8)

The fundamental solution of the Cauchy problem (12.3.8) at the temporal interval $\Delta t = t - \tau$ is expressed with the help of a function $f(\tau, \vec{p}, t, \vec{q})$:

$$\Omega(t, \vec{q}) = \iint_G \Omega(\tau, \vec{p}) f(\tau, \vec{p}, t, \vec{q}) d\xi d\eta,$$

$$\vec{q} = (x, y), \quad \vec{p} = (\xi, \eta).$$

(12.3.9)

For the function f is valid the equation

$$f(\tau, \vec{p}, t, \vec{q}) = \iint_G f(\tau, \vec{p}, \tau + \Delta\tau, \vec{z}) \cdot f(\tau + \Delta\tau, \vec{z}, t, \vec{q}) dz_1 dz_2, \quad (12.3.10)$$

which proves to be just the integral equation, connecting the values of the function $f(\tau, \vec{p}, t, \vec{q})$ at the different moments of time. It coincides in its form with the integral equation by Smoluchowski, which is well known in a statistical physics. Satisfied by the Smoluchowski equations are the densities of a probability of transition for such systems, which in their development pass through the sequence of states forming in their totality the Markovian chain. For the equation by Smoluchowski one might turn to that by Fokker–Plank, keeping oneself in assumption that the velocity field within the small interval of time is known. Thus one obtains

$$\frac{\partial f}{\partial t} + V_x \frac{\partial f}{\partial x} + V_y \frac{\partial f}{\partial y} = \nu \Delta f,$$

(12.3.11)

$$f(\tau, \vec{p}, t = \tau, \vec{q}) = \delta(\vec{q} - \vec{p}).$$

The method of solution of the Fokker–Plank equation are well developed in the theory of accidental processes, and their application to the solution of hydro-dynamical problems is, undoubtedly, rather interesting.

The set of stochastical differential equations, which are statistically equivalent to the Fokker–Plank equation, has a form

$$\frac{d\vec{q}_j(t)}{dt} = \vec{V}(t, \vec{q}_j) + \sqrt{2\nu}\vec{\xi}(\vec{q}_j, t),$$

$$\vec{q}_j(t = \tau) = \vec{p}_i(\tau), \quad i, j = 1, 2, 3 \ldots, \tag{12.3.12}$$

$$\vec{q}_j(t) = (x_j, y_j).$$

Here $\vec{\xi}(\vec{q}_j, t) = (\xi_x, \xi_y)$ is an accidental vectorial function, possessing the properties of a "white noise", $\vec{V}(t, \vec{q}_j)$ is a vector of the mathematical expectation of velocity (mean velocity) of a flow.

Using this method was solved a number of problems presented in Chaps. 5 and 7. Possibly, by the large Reynolds numbers the coincidence between the results obtained and those of the boundary-layer theory will be better. However, just as in the case of any other method based on the solution of the complete set of Navier–Stokes equations, by the large Reynolds numbers, for the achievement of a high accuracy this method will demand the quite considerable computational resources. Nevertheless, the originality of an approach, the importance and complexity of the hydrodynamical problems are, undoubtedly, making such an approach to be a prospective one, and, quite certainly, there will be found a class of problem, for the solution of which the method proposed will appear to be the optimally effective one.

12.4 Modeling of the Turbulent Flows of Fluid and Gas

The turbulence presents in itself the most complicated area of the mechanics of fluid and gas, connected with the stochastical processes, with the complicated nonlinear equations, with the multi-dimensionality, and with the large volumes of information. The numerical study of the various phenomena in turbulence is hampered by the fact that the existing mathematical models of turbulence are not numerous and not perfect. Moreover, many of these phenomena do not possess the reliable physical model. For these reasons, those properties of the Monte Carlo methods in their application to the study of turbulent flows, which were formulated in the Introduction and in Sec. 12.3 of the present Chapter, acquire the specific importance. Thus, in Secs. 10.3 and 10.4 of Chap. 10 two physical models are proposed:

(1) description of the small-scale turbulence;
(2) transition of the laminar boundary layer into the turbulent one.

In spite of the completely different physical nature of these models, their mathematical description is realized through the kinetically similar equations. One of such equations is the equation by Onufriev–Lundgren for the function of distribution of the velocities of pulsations, and another one — for the function of distribution of the density of wave packets within the space of wave vectors. And this situation permits to apply the methodics of solution of the kinetic equations, well elaborated in the dynamics of a rarefied gas, to the kinetically similar equations, obtained in Secs. 10.3 and 10.4 of Chap. 10.

One another, even more characteristical situation with the necessity of application of the Monte Carlo methods is revealed by the solution of the problem by Emmons — that of the transition of laminar boundary layer into the turbulent one (see Sec. 11.2 of Chap. 11). For this problem one not only has no mathematical model, but also is not quite clear the physical nature of the appearance and evolution of the "Emmons spots". There is only the wide experimental information on this subject. This material proves to be sufficient for the construction of a stochastical process of the modeling of the birth and evolution of the spots, and for the determination of the drag of a body located within the turbulent flow.

All these problems are connected with large volumes of the information and with labor-consuming computations, and for these reasons, of course, were developed the various methodics of the distribution of computations over the parallel processors.

12.4.1 *The fluidical model of turbulence*

In the proper time even Prandtl turned his attention to the fact, that there exists an analogy between the rarefied gas and the turbulent fluid. As a generalization of the application of kinetic models to the continuous media, in Sec. 10.4 of Chap. 10 was made an attempt to describe the turbulent phenomena. Here, just as it is in rarefied gas dynamics, the problem is solved at the level of a distribution function. But now the role of an argument is played not by the molecular velocity $\vec{\xi}$, but by the pulsation of the velocity of a liquid particle v. In this model each particle, located within the cell, possesses the new quality.

The liquid particle, just as it was before, is characterized by the physical coordinates and velocity. For the corresponding distribution function is proposed the model of a kinetic equation, similar to the model equation in the rarefied gas dynamics.

Rarefied gas dynamics	Turbulence
Particles	
Molecules	Liquid particles
r_i, coordinates of molecules	x_i, coordinates of particles
c_i, velocities of molecules	v_i, velocities of pulsations
Distribution function	
For the molecules	For the liquid particles
$f = f(t, r, c)$	$f = f(t, x, v)$
$\int f dc = \rho$ — density	$\int f dv = 1$ — rat setting
Moments	
$\frac{1}{\rho}\int c f dc = u$ — the macroscopic velocity,	$\int v f dv = u$ — the mean velocity
$(c - u)$ — the thermal velocity	$(v - u)$ — the fluctuations

Used for the description of a turbulence is the kinetic equation by Onufriev–Lundgren:

$$\frac{\partial f}{\partial t} + v\frac{\partial f}{\partial x} - \frac{1}{2\tau_1}\frac{\partial}{\partial v}(v'f) = \frac{f_M - f}{\tau_2},$$

where $f_M = \left(\frac{3}{4\pi E}\right)^{3/2}\exp\left[-\frac{3v'^2}{4E}\right]$ is a normal law and E is a turbulent density of the energy.

This kinetic equation is similar to the model equation by Krook, which was described in the Chap. 1.

Here the scheme of modeling is built up in accordance to the same principles, as in the rarefied gas dynamics. Considered are the liquid particles within the cells, and the whole process is divided into the three main steps:

1. The convective transfer

$$\left(v\frac{\partial f}{\partial x}\right).$$

2. The turbulent dissipation of the energy

$$-\frac{1}{2\tau_1}\frac{\partial}{\partial v}(v'f).$$

3. The redistribution of the energy

$$\frac{f_M - f}{\tau_2}.$$

Such an approach seems to be prospective for the modeling of small-scale turbulence. Using this method, were solved the problems of a dissipation of the turbulent spot and of an interferention of the turbulent spots.

12.4.2 Description of the turbulence with the help of a model of three-wave resonance

Considered in the approximation of a boundary layer are the equations by Reynolds for the mean values U and V, and also pulsations u, v, w, p:

$$\partial \bar{U}/\partial T + \bar{U}\partial \bar{U}/\partial X + \bar{V}\partial \bar{U}/\partial \bar{y} = -\partial \langle \overline{uv} \rangle /\partial \bar{y} + (1/\varepsilon^2 R)\partial^2 \bar{U}/\partial^2 \bar{U}\partial \bar{y}^2,$$

$$\partial \bar{U}/\partial X + \partial \bar{V}/\partial \bar{y} = 0, \quad R = U_\infty d/v,$$

$$\partial \bar{u}_i/\partial \bar{t} + \bar{U}\partial \bar{u}_i/\partial \bar{x}_1 + f_i = -\partial \bar{p}/\partial \bar{x}_i + (1/R)\nabla^2 \bar{u}_i + \varepsilon_i^T + o(\varepsilon^2),$$

$$\partial \bar{u}_i/\partial \bar{x}_i = 0, \quad R = U_\infty \delta^{**}/v,$$

$$i = 1, 2, 3.$$

By the proper assumptions (see Sec. 10.3 of Chap. 10) concerning the "density" of wave packets $I(k)$ is obtained the equation of transfer:

$$\partial I(\bar{k})/\partial T + g(\bar{k}, X)\partial I(\bar{k})/\partial X - h(\bar{k}, X)\partial I(\bar{k})/\partial \bar{k} - 2\bar{\omega}^I I(\bar{k}) = J_c,$$

$$g(\bar{k}, X) = \partial \bar{\omega}^R(\bar{k}, X)/\partial \alpha, \quad h(\bar{k}, X) = \partial \bar{\omega}^R(\bar{k}, X)/\partial X.$$

This equation resembles the kinetic one and describes the birth of the quasi-particles within the area of instability in respect of lower mode of the waves of Tollmien–Schlichting (T–S waves), their transfer with a group velocity $g(k, X)$ under the action of a "force" $(-h(k, X))$, as well as a disintegration and a confluence of the particles due to the three-wave resonance, which is described by a term J_c. In the elementary case the integral of collisions of that type appears as

$$J_c = \int d\bar{k}_1 P(\bar{k}, \bar{k}_1)\delta(\bar{\omega}^R(\bar{k}) - \bar{\omega}^R(\bar{k}_1) - \bar{\omega}^R(\bar{k}_2))(I(\bar{k}_1)I(\bar{k}_2)$$

$$- I(\bar{k})I(\bar{k}_1) - I(\bar{k})I(\bar{k}_2)),$$

$$\bar{k} = \bar{k}_1 + \bar{k}_2.$$

To study the equation for a wave packet $I(k)$, which is similar to the kinetic equation, might be, in principle, applied the method of a direct statistical modeling, which has well recommended itself by the modeling of phenomena described by the kinetic equations. For example, one might propose the following methodics of the direct statistical modeling (Monte Carlo). Let us look for a solution of the equation $I(k)$ for within the interval Δt, assuming that Δt is a small quantity. Such a solution of the equation for $I(k)$, during the time Δt, might be written down as follows:

$$I_{\bar{k}}(t + \Delta t) = I_{\bar{k}}(t)(1 - A_k \Delta t) + A_k \Delta t \left(\gamma'_k + \int (k' I_{k_1} I_{k_2}) d\bar{k}_1 \right),$$

$$A_{\bar{k}} = \int V_{kk_1 k_2} (I_{k_1} + I_{k_2}) d\bar{k}_1,$$

$$k_{\bar{k}\bar{k}_1\bar{k}_2} = V_{\bar{k}\bar{k}_1\bar{k}_2} \delta(\omega^R(\bar{k}) - \omega^R(k_1) - \omega^R(k_2)),$$

$$k' = k_{\bar{k}\bar{k}_1\bar{k}_2}/A,$$

$$\gamma'_{\bar{k}} = \gamma_k/A.$$

Let us treat the function $I_{\bar{k}}$ as a density of certain particles within the space of the vectors $\bar{k}$. This density varies as the time goes on, and the connection of the value of it at the moment of time $t + \Delta t$ with the value at the moment of time t is given by the preceding equation.

Since it is assumed that the function I_k at the moment of time t is known, then the quantities $A_{\bar{k}}$, $\gamma'_{\bar{k}}$ and k' are known, too. If the value Δt is chosen in such a way, as to satisfy the inequality $A_k \Delta t < 1$, then it would be possible to use the principle of superposition. This means that the value of density with a probability $(1 - A_{\bar{k}} \Delta t)$ at the moment of time $t + \Delta t$ is equal to $I_{\bar{k}}(t)$, while with a probability $A_{\bar{k}} \Delta t$ it is equal to $(\gamma'_{\bar{k}} + \int k' I_{k_1} I_{k_2} d\bar{k}_1)$. If this density at the moment of time t is modeled by the certain number of particles (at the initial moment of time this number might be eqal to zero), then with a probability $(1 - A_{\bar{k}} \Delta t)$ these particles are not changing their coordinates, while with a probability $A_{\bar{k}} \Delta t$ either will be born the new particles with a frequency k', or the coordinates of particles are changing with a frequency $\gamma'_{\bar{k}}$. The expression, connected with the birth of particles, might be considered as a superposition:

$$\gamma'_{\bar{k}} + \int k' I_{\bar{k}_1} I_{\bar{k}_2} d\bar{k}_1 = C_1 \gamma'_{\bar{k}} + C_2 \int k' I_{\bar{k}_1} I_{\bar{k}_2} d\bar{k}_1,$$

$$C_1 = \frac{1}{C} \int \gamma'_{\bar{k}} d\bar{k}, \quad C_2 = \frac{1}{C} \int k' I_{\bar{k}_1} I_{\bar{k}_2} dk \, dk_1,$$

$$C = C_1 + C_2.$$

The algorithm of computation of $I_k(t + \Delta t)$ looks as follows:

1. In the space of vectors $\bar{k}$ are prescribed the coordinates of a number of particles in accordance to the initial density (which might be equal to zero, and in this case there are no particles within a space).
2. Computed are $A_{\bar{k}}$, k', $\gamma'_{\bar{k}}$. If it proves to be, that $A_{\bar{k}} \Delta t > 1$, then Δt is changed.
3. With a probability $(1 - A_{\bar{k}} \Delta t)$ the field is not changed, and after the calculation of the necessary macroquantities is realized transfer to the next step (see point 2).

4. With a probability $A_{\bar{k}} \Delta t$ occurs the birth of particles.

    4.1 With a probability C_1 the particles are born with a density γ'.

    4.2 With a probability C_2 the particles in a field are changing their coordinates.

    4.3 After reaching the boundary of a computational area the particles are vanishing.

After the calculation of the necessary macroquantities is realized a transfer to the next step.

The experience in using the reliable methodics of the statistical modeling of flows and of the solution of kinetic equations (for example, the method by Belotserkovskii–Yanitskii) in the area of continuous medium and of turbulence (Chaps. 4, 5, 10 and 11) permits to hope that the proposed here method of the modeling of a turbulent transition will prove to be widely used in the further practice.

12.5 Parallelization of the Statistical Algorithms (Bukin, Voronich, Shtarkin)

The parallelization of computations for the high-productive supercomputer systems appears to be one of the main ways of development of the modern computational mathematics. The supercomputers are the more and more widely used for a solution of the fundamental and applied problems in the areas of nuclear physics, climatology, economics, pharmacology, modeling of the training devices, and of the virtual reality, computational aerodynamics. Due to those specific features of the Monte Carlo methods, which were repeatedly stressed in the present book, the statistical modeling begins to play the more and more noticeable role in all, indicated above areas of science and techniques. For these reasons, the actuality of the problems mentioned is growing very considerably, taking into account the fact that the computational aerodynamics is the most promoted area of the elaboration, development, and application of the Monte Carlo methods. As the mentioned above features of these methods (see point 2 of Sec. 12.3 of the present Chapter concerning the simple structure of a computational algorithm) permit to state, that the numerical schemes of a statistical modeling might be, in quite a natural way, transferred onto the parallel processors. The present authors are not aware of any studies on the parallelization of computations involving the methods, based on the modeling of trajectories of the "trial particles", but, nevertheless, the way of distribution of computations is, in this case, quite evident. Clearly, the successive modeling of the independent trajectories should be entrusted to the individual processors, while the information for the averaging will be gathered by a server.

Equally clearly is, that in this case, the productivity of the method is growing in direct proportionality to the number of parallel processors. As it is seen from the analysis of materials of the International Symposia on Rarefied Gas Dynamics, in the present time the statistical method, most popular all around the world, is the method of modeling of the "evolution of the ensemble of particles", proposed by Bird, and the modifications of that method.[9,10,45,49] Therefore, the most of scientific works on the distribution of computations is devoted just to that method. One of the ways of such a distribution is, moreover, sufficiently evident. Since the structure of a computational algorithm possess a similarity for any temporal cut, at each of the processors the problem is set in its complete volume, while the information for averaging in time is coming to a server. Once again, it is evident, that the productivity of a method grows in a direct proportionality to the number of parallel processors. With the help of this method was solved, in particular, the rather labor-consuming problem of the interferention of two turbulent spots (see Sec. 11.2 of Chap. 11) (Figs.12.3 and 12.4).

Clearly, such an evident way of parallelization of the computations lays rather essential limitations on the class of those problems, which might be treated with the help of parallels computers. First, each individual processor should possess a

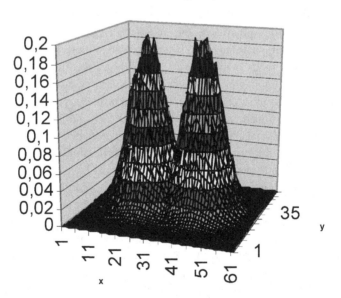

Fig. 12.3   Functions of the distribution of turbulent energy of the interacting spots at $t = 0$.

t = 40

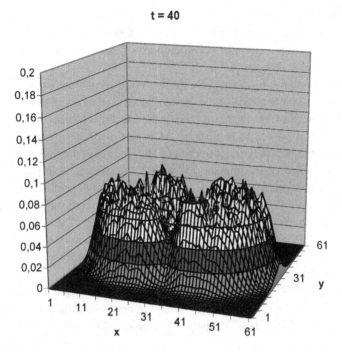

Fig. 12.4   Functions of the distribution of turbulent energy of the interacting spots at $t = 40$.

sufficient power for the autonomous accommodation inside of itself of the complete problem, and such a demand is automatically excluding the modeling of complicated and multi-dimensional processes with a large volume of information. Second, lost is the possibility of using the memory of the parallel system in its totality, as well as its additional possibilities. And there is yet another reason, which stipulates the necessity for the elaboration of new methods of the distribution of computations.

The supercomputers are rather expensive and not easily accessible. In the quality of their alternative appear the network cluster systems, or the network clusters, that is, the totalities of computers united by a network and possessing the common control. The network cluster systems are comparatively inexpensive, relatively simple in their arrangement, and by the sufficient quantity of active machinery they provide the productivity, comparable to that of supercomputers. For the realization of the common control for the network cluster system it would be necessary to have the special program provision, as, for example, MPI,[226–228] PVM,[229] mpC,[230] Linda,[231] T-system.[232] Just the program provision defines the special architectural features of the network cluster, as a powerful computational complex.

Used here is the library of Message Passing Interface (MPI), which presents a low-level, but, at the same time, extremely convenient interface of the programming for a network cluster, and is based on the idea of exchange by communications among the parallel processors.

The programming for a network clusters is different from the usual models of programming on the basis of one processor, or even of the multitude of them. As concerns the network, the realization of the usual mechanism of the exchange of information is proved to be difficult because of the high overhead expenses, stipulated by the necessity to let to each of the processors the individual copy of one and the same dividable memory. Moreover, the solution of modern problems demands the increase of a total volume of the accessible operative memory of parallel computers. For this reason, by the programming for the network clusters is used, as a rule, the Single Program-Multiple Data technology (SPMD)[233]. The idea of SPMD consists in a tendency to divide the large array of information between the identical processors. After that each of the processors will carry out the processing of its part of data (see Fig. 12.5).

In the case of SPMD-approach it would be sufficient to send from time to time to the processors the blocks of data, which are demanding the labor-consuming treatment, and afterwards — to gather the results of their work. If the time of

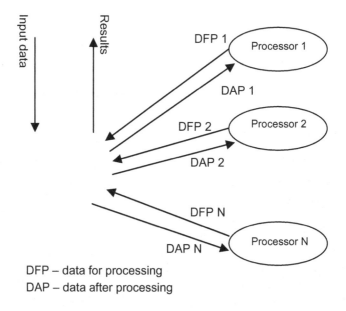

DFP – data for processing
DAP – data after processing

Fig. 12.5

processing of the block of data by one processor is considerably larger than the time of sending of that block through the network, then the network cluster systems becomes to be very effective. Just such an approach is used in MPI. Here always exists a certain main processor, which realizes the distribution of data between the other processors, and upon the termination of computations gathers the results and demonstrates them to the user. Usually, after the distribution of data the master processor carries out the processing of the part of these data, thus aspiring to use the system's resources in the most effective way. Actually, each of the communications presents in itself a packet of the typified data, which one of the processors might send to the other one or to the group of them.

12.5.1 *Structure of the parallel algorithm*

Here in the basis of a parallel algorithm is set the method by Belotserkovskii–Yanitskii. The algorithm is elaborated taking into account the demand for the low expenditures on the acception/transmission and on the processing of information connected with the interaction of the master processor and the several subordinated processors. Thus, the structure of algorithm appears in the following form.

1. The motion of particles is divided into two stages: collision and transfer.
2. Before the beginning of the calculations each of the subordinated processors obtains from the master one the range of those numbers of cells (the numeration of the cells is one-dimensional and through), which this processor should use, as well as the initial data introduced by user. After that each of the subordinated processors generates the particles within its own range of cells and in accordance to the initial conditions.
3. Begins the process of computation, by which each of the subordinated processors works only with its range of cells.
4. Before each of the new steps is carried out the computation of dt-step in time. Each of the subordinated processors sends its minimal dt to the master one, where is determined the common minimal dt and sent to all the subordinated processors.
5. Each of the subordinated processors realizes the collision of particles.
6. Each of the subordinated processors realizes the transfer of particles. During that event those particles, which fly out of the range of cell's numbers for the present processor, are transported into the buffer for the subsequent sending to the master processor.
7. Sending of the particles, which flew out of the range of cell's numbers to the master processor.

8. Sorting of the particles accepted into groups connected with processors and sending these particles to the subordinated processors.
9. Repeating of the points from 4 up to 8, till the moment, when will be reached the prescribed time *t* of the experiment.

The scheme of work of the parallel algorithm with a designation of the stages of the data exchange between the processors.

Master processor	Subordinated processor
1. Introduction of the parameters of computation	1.
2. Sending of the parameters of computation to the subordinated processor	2. Obtainment of the parameters of computation
3. Transference of the boundaries of a computational area	3. Obtainment of the boundaries of a computational area
4.	4. Initialization of the initial distribution
5. Obtainment and the choice of minimal step *dt*	5. Transference of the step *dt*
6. Transference of the minimal temporal step *dt*	6. Obtainment and recording of step *dt*
7.	7. The cycle over all the cells: • Collision of the particles • Transfer of the particles • The particle in computational area → array of the particles • The particle out of computational area → buffer for the transference
8. Obtainment of the arrays of particles	8. Transference of buffer to the master processor
9. Formation of arrays for subordinate processors	9.
10. Transfer of the arrays to the corresponding subordinate processors	10. Obtainment of the array of particles caught by the area of computations for the present computer
Repetition of the steps from 5 to 10	Repetition of the steps from 5 to 10

The method proposed here assumes the unique realization with the use of a large number of particles and obtainment of the necessary statistics with the help of one and only computation. It is also assumed, however, that with the present method the complete array of particles is not kept within any of the processors. This situation permits to sum up the volumes of an operative memory and to compute the problems with a high demands in respect to the number of particles. Moreover, the transfer of information by large packets is increasing the effectivity of the algorithm. The approbation of a method was conducted in application to the problem of modeling of turbulence in the rarefied gas (see Sec. 11.3 of Chap. 11) (Fig. 12.6).

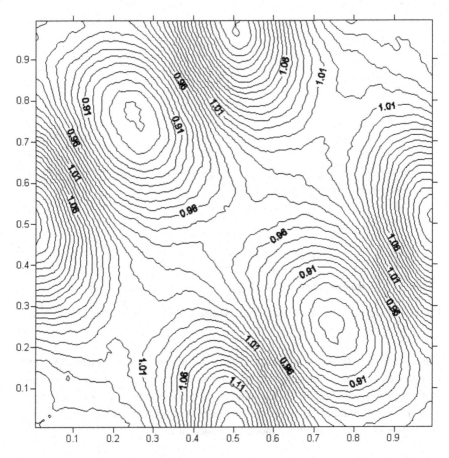

Fig. 12.6   Evolution of the vortical system (densities) at $t = 0.1$.

The results of application of the present method are identical to those described in Ref. 222, but the effectivity of the algorithm proposed is in several times higher (see. Fig. 12.7).

For a unit of velocity in Fig. 12.7 is taken that of a work of the present program in the regime — one master processor plus one subordinated processor, which is equivalent to the one-processor regime.

The results obtained indicate to the essential increase of the computational speed by the increase of a number of active knots in a cluster. This tendency is stipulated both by the low expenditures for a transfer of computational data, and by the effective realization of the principle itself of the parallelization proposed.

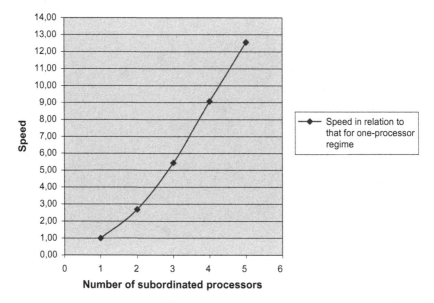

Fig. 12.7   Diagram of the dependence of computational speed on the number of subordinated processors.

Namely, the use of a notion of the distributed operative memory, when the operative memory of several computers is acting just as an integrity. Such an approach is increasing not only the accessible volume of a memory, but also its aggregate working ability, since the processes of reading/writing are going on at several computers in parallel and independently of each other.

Conclusions

Thus, in the present monograph is carried out the analysis of Monte Carlo methods, elaborated in the rarefied gas dynamics, and are presented the results of studies, obtained with the help of these methods within the wide range of the degrees of rarefaction of a medium, of the velocities, of the temperatures, of the geometrical forms, and of the other parameters of flow. The demonstrative role is played by the history of development of the Monte Carlo methods, connected with a creation of the atomic technique and of the space-rocket one. Are stressed the main properties of the methods, which are explaining their permanently increasing part in the solution of fundamental and applied problems in the various areas of physics, climatology, biology, and economics.

In particular, side by side with a "heavy" dependence of the error of a method on the number of trials N:

$$I = \frac{1}{N} \sum_{i=1}^{N} \xi_i \pm \frac{3\sigma}{\sqrt{N}},$$

which, as it was shown earlier, might be successfully overcome by way of using of the physical nature of a process modeled, the Monte Carlo methods possess the complete set of the remarkable peculiarities, which reveal their profitable distinction from the regular numerical schemes:

(1) the error has not, practically, any connection with the dimensionality of the problem;

(2) the simple structure of a computational algorithm (N times repeated and belonging to one type calculations of the realization of accidental quantity or accidental trajectory);

(3) the construction of an accidental quantity ξ might be, in general case, based on the physical nature of a process, and will not demand the obligatory, as in the regular methods, formulation of the equation, which property becomes to be more and more actual for the study of modern problems;

(4) the united statistical procedure permits to build up the general approach to the direct modeling of the processes of various physical nature;

253

(5) the single-typed structure of a computational algorithm permits to find the natural way to a construction of the effective schemes of parallel computations.

In this connection is growing the actuality of the studies conducted, since the computational aerodynamics looks as the most advanced area of physics, as concerns the elaboration, the substantiation and the application of Monte Carlo methods. Thus, formulated is the connection between the direct statistical modeling of the aerodynamical processes and the solution of kinetic equations, and it is shown that the contemporary stage of the development of computational methods proves to be inconceivable without a complex approach to the construction of algorithms taking into account all the peculiarities of the problem to be solved: the physical nature of a process, the mathematical model, the theoretical aspects of computational mathematics, and of stochastical processes. Considered are the possible ways of development of the methods of statistical modeling. Specially noted is the universality of a statistical approach to the modeling of the flows of fluid and gas, including the sufficiently complicated ones. Studied are the new possibilities of Monte Carlo methods in their application to the adjacent, nontraditional concerning the use of a statistical modeling, areas, like the research of the flows of continuous medium (equations by Euler and by Navier–Stokes) and the study of turbulence. The monograph is based, mostly, on the original results, obtained by its authors, as well on the lecture courses, presented by the authors in the Moscow Physico-Technical Institute.

As it was already noted in the Introduction, the book was created within the frame of science of a scientific project "POISK" ("THE SEARCH"), elaborated at the department of aeromechanics and flying technics of MPTI, of which the sense consists in the creation an original guide of one of the most complicated problems of the contemporary fundamental and applied science — that of the nonuniform and anisotropical turbulence. All around the world an enormous army of the research workers is busy with the solution of that problem.

Presently, the terrific amount of the actual materials is accumulated, and it becomes the more and more difficult to orient in that material without the proper guide, especially, as concerns the young people. The structure of a project mentioned presents in itself the creation of book with the analysis of experimental results, of the theoretical and computational methods, and with indication of the main directions of a research. The project is already partly realized. In particular, is published the book with a survey of the modern experimental studies on the dynamical structures in the turbulent boundary layer:

Yu.I. Khlopkov, V.A. Zharov, S.L. Gorelov. The Coherent Structures in the Turbulent Boundary Layer. Moscow, MPTI, 2002.

In this book, containing over 400 references, are presented the main features of the physics of dynamical processes in the turbulent boundary layer, such as the phenomenon of bursting, formation of the streaks, processes of a transfer of momentum and energy from the outer boundary of a boundary layer to the boundary of a flow. Moreover, is presented a critical analysis of the native and foreign experimental studies, are formulated the actual problems. It was found that the analysis of experimental works conducted within a lengthy period (over 40 years) revealed certain essential peculiarities of the flows of fluid and gas, which might be used by the construction of the general theory of such processes and by the creation of the physical models of turbulence.

The theoretical investigations of turbulent flows are also conducted during a long time. The considerable part of that time was devoted to a search of the most effective methods of solution of the corresponding problems. In the survey book Yu.I. Khlopkov, V.A. Zharov, S.L. Gorelov, Lectures on the Theoretical Methods of the Investigation of Turbulence. Moscow, MPTI, 2005, 180 pp., were summed up the results of these investigations, was presented the critics of various methods, which were used at the earlier stage of the development of theory. All this permits one to orient in the contemporary directions of investigation.

Not long time ago by the Editorial and Publishing Department of MPTI was published the survey book: Yu.I. Khlopkov, V.A. Zharov, S.L. Gorelov, Renormgroup Methods of Description of the Turbulent Motions of Incompressible Fluid. Moscow, MPTI, 206, 492 pp. Presented in that book is the survey of results of the elaboration and application of the number of such methods, which were nominated as renormgroup ones, to the construction of models of the turbulent flows of incompressible fluid both for the uniform and isotropic case, and for the case of a strong anisotropy and nonuniformity. The book is based on the study of several hundreds of the original works, from the totality of which were chosen the most actual ones, on the authors opinion. The largest part of a book is devoted to the three sub-network models of turbulence, which are widely used in the contemporary practical activity of various experts in the field of aerodynamics. The book is published as a textbook for students in spite of the fact that it demands the considerable efforts for its understanding, and, actually, is intended for the professors and postgraduates.

At the present time is prepared for publication "The Course of Lectures on the Theory of Turbulence", and these lectures were presented at the department of aeromechanics and flying technics of MPTI by the Professor V.N. Zhigulev, the well-known scientist, who for a rather long time was busy with investigations in this area and has deeply penetrated into the essence of a problem at the kinetic level.

Further on, it is planned to carry out the survey and analysis of the modern numerical methods, which are applied to the modeling of the complicated unsteady flows of fluid and gas. Such is the general plan of work on the creation of a scientific basis and guide. It is necessary to note that the work on the creation of such a guide is extremely complicated and labor-consuming. At the present moments this project contains more than 1000 references. And the authors are expressing their deepest gratitude to the Russian Fund of the Fundamental Research, which is stably supporting the present Project.

References

[1] V. S. Vladimirov and I. M. Sobol, Calculation of the minimal characteristical number of the Pyerls equation by Monte Carlo Method, *Comput. Math.* (3), 1958 (in Russian).

[2] S. M. Ermakov, *Monte Carlo Method and Adjacent Questions* (M., Science, 1958) (in Russian).

[3] M. Kac, *Probability and Adjacent Questions* (M., Mir, 1965) (in Russian).

[4] M. N. Kogan, *Rarefied Gas Dynamics* (M., Science, 1967) (in Russian).

[5] O. M. Belotserkovskii and V. E. Yanitskii, Numerical methods in dynamics of a rarefied gas, II, *Proc. of the 4th All-Union Conf. on Raref. Gas Dyn*, AS USSR, TsAGI (M., 1975) (in Russian).

[6] *Computational Methods in Rarefied Gas Dynamics* (M., Mir, 1969) (in Russian).

[7] G. Bird, *Molecular Gas Dynamics* (M., Mir, 1981) (in Russian).

[8] O. M. Belotserkovskii, *Numerical Modeling in Mechanics of Continuous Media* (M., Science, 1984) (in Russian).

[9] O. S. Ryzhov, Numerical methods in dynamics of rarefied gases, *Development and Use at the Comp. Center of Acad. of Sci. of USSR*, CC AS USSR (1977) (in Russian).

[10] M. A. Ender and A. Y. Ender, On one presentation of the Boltzmann equation, *Acad. of Sci. USSR Dokl.* **193**(1), 61–64 (1970) (in Russian).

[11] F. G. Tcheremissin, Numerical solution of the kinetic equation by Boltzmann for one-dimensional steady motions of a gas, *J. Comp. Math. & Math. Phys.* **6**(3), 654–665 (1970) (in Russian).

[12] F. G. Tcheremissin, Development of the method of direct numerical solution of the Boltzmann equation, *Num. Meth. in Dyn. of Raref. Gases*, CC AS USSR, M. (1), 74–101 (in Russian).

[13] S. L. Shcherbak, On the solution of a problem of flow about the half-infinite plate on the basis of Boltzmann equation, *Proc. of the High Aviat. School of Civic Aviation*, Leningrad, 45th issue, pp. 96–109 (1970) (in Russian).

[14] E. M. Shakhov, *Method of Study of the Motions of Rarefied Gas* (M., Science, 1974) (in Russian).

[15] E. M. Shakhov, The crosswise flow about the plate by a rarefied gas, *AS Izv., Mech. of Fl. Gas* (6), 107–113 (1972) (in Russian).

[16] I. N. Larina, Flow of rarefied gas about a sphere, *Appl. Math. & Mech.* **33**(5) (1969) (in Russian).

[17] I. N. Larina and V. A. Rykov, Aerodynamics of a sphere moving in the flow of a rarefied gas, *AS USSR Izv., Mech. of Fl. Gas* (3), 173–176 (1983) (in Russian).

[18] E. F. Limar, Numerical study of the flow of rarefied gas about a cylinder, *Num. Meth. in Dyn. of Raref. Gases*, CC AS USSR (2), 95–107 (1975) (in Russian).

[19] I. M. Sobol, *Numerical Methods of Monte Carlo* (M., Science, 1973) (in Russian).

[20] S. M. Ermakov and G. A. Mikhailov, *Course of the Statistical Modeling* (M., Science, 1976) (in Russian).

[21] G. I. Marchuk, G. A. Mikhailov, *et al.*, *Monte Carlo Method in Atmospheric Optics* (Novosibirsk, Science, 1976) (in Russian).

[22] Y. I. Khlopkov, Solution of the linearized Boltzmann equation, *J. Comp. Math. Math. Phys.* (5) (1973) (in Russian).

[23] S. N. Ermakov, On the analog of the scheme by Neumann–Uhlam in the nonlinear case, *J. Comp. Math. Math. Phys.* (3) (1973) (in Russian).

[24] S. N. Ermakov and Nefedov, On the estimates of the Neumann's sum by Monte Carlo method, *AS USSR Dokl.* **202**(1), 27–29 (1972) (in Russian).

[25] Y. I. Khlopkov, Statistical method of solution of the approximate kinetic equation, *TsAGI Scient. Notes* **4**(4), 108–113 (1973) (in Russian).

[26] Y. N. Grigoriev, M. S. Ivanov and M. I. Kharitonova, On the solution of the nonlinear equations of rarefied gas dynamics by Monte Carlo method, *Num. Meth. of Mech. Cont. Med.*, AS USSR, Siber. Branch, CC, Novosibrsk, **2** (1971) (in Russian).

[27] V. I. Vlasov, Improvement of the method of statistical trials (Monte Carlo) for computation of the flows of rarefied gas, *AS USSR, Dokl.* **167**(5) (1966) (in Russian).

[28] V. I. Vlasov, Computation by Monte Carlo method of the heat flow between the parallel plates in rarefied gas, *TsAGI Scient. Notes* **1**(4), 46–51 (1970) (in Russian).

[29] A. I. Eropheev and V. A. Perepukhov, Computation of a flow about the plate, located along the flow of rarefied gas, *TsAGI Scient. Notes* **6**(3), 51–57 (1975) (in Russian).

[30] V. E. Yanitskii, Application of the stochastic process by Poisson for computation of a collisional relaxation of non-equilibrium gas, *J. Comp. Math. Math. Phys.* **13**(2), 505–510 (1973) (in Russian).

[31] O. M. Belotserkovskii and V. E. Yanitskii, Statistical method of "particles in cells" for solution of the problems of rarefied gas dynamics, *J. Comp. Math. Math. Phys.* **15**(5), 6 (1975) (in Russian).

[32] O. M. Belotserkovskii and V. E. Yanitskii, Numerical methods in rarefied gas dynamics, *Proc. of the 4th All-Union Conf. on Raref. Gas Dyn. and Molecular Gas Dyn.* M., TsAGI, 101–183 (1977) (in Russian).

[33] O. M. Belotserkovskii and V. E. Yanitskii, Direct numerical modeling of the flows of rarefied gas, *AS USSR, CC., M.* (3), 81–88 (1977) (in Russian).

[34] V. E. Yanitskii, Application of some statistical models to a numerical solution of Boltzmann equation, *Master of Sci. Diss.*, CC AS USSR (1974) (in Russian).

[35] V. E. Yanitskii, Application of the processes of accidental wandering, *J. Comp. Math. Math. Phys.* (1) (1974) (in Russian).

[36] V. V. Sychev, The Asymptotical Theory of Breakdown Flows, *AS USSR IZV., Mech. of Fl. Gas* (2) (1982) (in Russian).

[37] V. A. Perepukhov, Aerodynamical characteristics of a sphere and of a blunted cone in the flow of strongly rarefied gas, *J. Comp. Math. Math. Phys.* **7**(2) (1967) (in Russian).

[38] V. A. Perepukhov, Application of the Monte Carlo method in dynamics of the strongly rarefied gas, *TsAGI Proc.* (1411) (1972) (in Russian).

[39] A. I. Eropheev and V. A. Perepukhov, Computation of the crosswise flow about a plate by the rarefied gas, *AS USSR Izv., Mech. of Fluid and Gas* (4), 106–112 (1976) (in Russian).

[40] S. L. Gorelov and A. I. Eropheev, The influence of internal degrees of freedom on the hypersonic flow of rarefied gas about a plate, *AS USSR Izv., Mech. of Fluid and Gas* (6), 151–156 (1978) (in Russian).

[41] V. I. Vlasov, Computation of the aerodynamical characteristics of the plane plate of infinite span in the hypersonic flow of rarefied gas, *TsAGI Scien. Notes* **2**(6), 116–118 (1971) (in Russian).

[42] V. I. Vlasov, Computation of the flow of rarefied gas about the plate at the angle of attack, *TsAGI Scien. Notes* **4**(1), 17–24 (1973) (in Russian).

[43] V. I. Vlasov, Computation by the Monte Carlo method of the rarefied gas flow about a plate at the angle of attack, *Proc. of 4th All-Union Conf. on Raref. Gas Dyn. and Molec. Gas Dyn. M., TsAGI*, 353–357 (1977) (in Russian).

[44] A. I. Eropheev, On the modeling of intermolecular interaction by the solution of Boltzmann equation by Monte Carlo method, *AS USSR Izv., Mech. of Fluid Gas* (6), 171–174 (1977) (in Russian).

[45] A. I. Eropheev and V. A. Perepukhov, Flow of a rarefied gas about the plate, *Proc. of 4th All-Union Conf. on Raref. Gas Dyn. Molec. Gas Dyn. M., TsAGI*, 358–364 (1977) (in Russian).

[46] V. I. Vlasov, A. I. Eropheev and V. A. Perepukhov, Computation of a flow of a rarefied gas about the plate, *TsAGI Proc.* (1830), 40 (1974) (in Russian).

[47] Y. N. Grigorjev and M. S. Ivanov, The flow of rarefied gas about a cylinder in transitional regime, *Num. Meth. of Cont. Medium*, AS USSR, Siberian Branch, CC Izv. **5**(1), 152–156 (1974) (in Russian).

[48] M. S. Ivanov, Solution of the axisymmetrical problems of rarefied gas dynamics by Monte Carlo method, *Proc. of the 4th All-Union Conf. on Raref. Gas Dyn. and Molec. Gas Dyn. M., TsAGI*, 388–391 (1977) (in Russian).

[49] Y. N. Grigorjev and M. S. Ivanov, On the solution of problems of rarefied gas dynamics by Monte Carlo method, *Appl. Aerodyn. of Space Apparatuses* (Kiev, Naukova Dumka).

[50] Y. I. Khlopkov, Wedge in a flow of rarefied gas, *TsAGI Scient. Notes* **7**(4) (1976) (in Russian).

[51] Y. I. Khlopkov, The drag of a sphere in the low-speed flow of a rarefied gas, *TsAGI Scient. Notes* **6**(5) (1975) (in Russian).

[52] Y. I. Khlopkov and E. M. Shakhov, Kinetic models and their role in studies of the flows of rarefied gas, *AS USSR Comp. Center, M.* (3) (1974) (in Russian).

[53] Y. I. Khlopkov, The hypersonic flow of rarefied gas about the axisymmetrical body, *TsAGI Scient. Notes* **9** (1978) (in Russian).

[54] Y. I. Khlopkov, The characteristics of a flow about a sphere by the super- and hypersonic speeds, *AS USSR Izv., Mech. of Fluid Gas* (3) (1981) (in Russian).

[55] Y. I. Khlopkov, The characteristics of a flow about a cone in traditional regime, at zero angle of attack, *AS USSR Izv., Meth. of Fluid Gas* (3) (1981) (in Russian).

[56] V. V. Serov and Y. I. Khlopkov, The improvement of the method of a direct non-steady modeling of a flow in dynamics of a rarefied gas, Theses of Papers, *9th All-Union Conf. on Raref. Gas Dyn.*, Sverdlovsk (1987) (in Russian).

[57] A. I. Eropheev, A. I. Omelik and Y. I. Khlopkov, Numerical and experimental modeling of the aerodynamical characteristics at high altitudes, *TsAGI, Competition for the Best Work* (1978) (in Russian).

[58] A. I. Eropheev, Three-dimensional hypersonic flow of a rarefied gas about a plate, *TsAGI Scient. Notes* **2**(5) (1978) (in Russian).

[59] A. S. Kravchuk, V. V. Serov and Y. I. Khlopkov, The possibilities of methods of the direct statistical modeling, *Celebrat. Book for the Jubilee of TsAGI, M.* (1990) (in Russian).

[60] S. L. Gorelov, A. I. Eropheev and Y. I. Khlopkov, *Numerical Modeling of the Aerodynamical Processes at High Altitudes* (Gagarin Readings. M., N., 1987) (in Russian).

[61] S. L. Gorelov and A. I. Eropheev, Flow of a diatomic rarefied gas about a cone, *TsAGI Scient. Notes* **15**(1) (1984) (in Russian).

[62] Y. I. Khlopkov, Statistical method of a solution of problems in gas dynamics, *Proc. of 6th All-Union Conf. of Raref. Gas Dyn.*, Novosibirsk (1980) (in Russian).

[63] A. M. Bishaev and V. A. Rykov, Solution of the stationary problems of the kinetic theory of gases by the moderate and small Knudsen numbers by the method of iterations, *Num. Meth. Raref. Gas Dyn.*, Issue (2), CC AS USSR, M. (1975) (in Russian).

[64] M. N. Kogan, A. S. Kravchuk and Y. I. Khlopkov, Method of "relaxation-transfer" for solution of the problems of gasdynamics in a wide range of medium's rarefaction, *TsAGI Scient. Notes* (2) (1988) (in Russian).

[65] E. V. Alekseeva and R. G. Barantsev, *The Local Method of the Aerodynamical Computation in Rarefied Gas* (M., LSU Publ. House, 1976) (in Russian).

[66] V. S. Galkin, A. I. Eropheev and A. I. Tolstykh, The approximate method of computation of the aerodynamical characteristics of bodies in hypersonic rarefied gas, *TsAGI Proc.* Issue 1833 (1977) (in Russian).

[67] A. I. Bunimovich and V. G. Chistolinov, The analytical method of computation of aerodynamical characteristics of bodies in the hypersonic flow of a gas of various rarefaction, *TsAGI Proc.* Issue 1833 (1977) (in Russian).

[68] V. Y. Ponomarjov and V. S. Seregin, Computation on the basis of local interaction of aerodynamical characteristics of complicated bodies by their stationary and non-stationary motion in rarefied gas, *Proc., 4th All-Union Conf. on Raref. Gas Dyn. and Molec. Gas Dyn., M., TsAGI* (1977) (in Russian).

[69] M. A. Zakirov, Study of the internal and external free-molecular flows near the arbitrary group of complicated bodies, *TsAGI Proc.*, Issue 1411 (1972) (in Russian).

[70] V. P. Bass, Computation of the flows of strongly rarefied gas taking into account the interaction with a surface, *AS USSR Izv., Mech. of Fl. Gas* (5) (1978) (in Russian).

[71] V. N. Kovtunenko, V. F. Kameko and E. P. Yaskevich, *Aerodynamics of the Orbutal Space Apparatuses* (Kiev, Naukova Dumka, 1977) (in Russian).

[72] R. G. Barantsev, *The Version of Local Method for Thin Bodies in Rarefied Gas* (Leningrad, LGU, 1982) Issue 13 (in Russian).

[73] R. G. Barantsev, The local theory of transfer of momentum and energy to a space in rarefied gas, *Math. Models, Anal. Num. Meth. in Theory of Transfer* (Minsk, 1982) (in Russian).

[74] M. A. Zakirov, A. I. Omelik and Y. I. Khlopkov, Theoretical and experimental investigation of the aerodynamical characteristics of simple bodies in hypersonic and

free-molecular flow, Novosibirsk, *6th All-Union Conf. of Raref. Gas Dyn.* (1979) (in Russian).

[75] Y. I. Khlopkov, Methodics and computer program for calculation of the characteristics of flying apparatuses in free-molecular regime, *TsAGI Proc.*, Issue 2111 (1981) (in Russian).

[76] E. V. Eremeev and Y. I. Khlopkov, *Engineering Methodics of a Computer Calculation of Aerodynamical Characteristics of Bodies of Complicated Form by the Flight in Transitional regime* (Interdepartment Collection, M., MPTI, 1988) (in Russian).

[77] E. V. Eremeev and Y. I. Khlopkov, Improvement of the engineering methodics of computation of aerodynamical characteristics of bodies of complicated form in the transitional regime, *Proc. of 33th Scient. Conf. of MPTI, M.* (November 1988) (in Russian).

[78] V. I. Vlasov, S. L. Gorelov and M. N. Kogan, Mathematical experiment for the computation of the coefficients of transfer, *AS USSR Dokl.* **176** (1968) (in Russian).

[79] Y. I. Khlopkov, Computation of the coefficients of transfer and of slip velocity for the molecules as solid spheres, *AS USSR Izv., Mech. of Fl. Gas.* (2) (1971) (in Russian).

[80] V. I. Vlasov and Y. I. Khlopkov, Version of Monte Carlo method for solution of linear problems of rarefied gas dynamics, *M., J. Comp. Math. and Math. Phys.* (4) (1973) (in Russian).

[81] M. S. Ivanov, Statistical modeling of the hypersonic flows of rarefied gas, Novosibirsk, Ph. D. Dissertation (1992) (in Russian).

[82] Y. I. Khlopkov, Ph. D. Dissertation (1998) (in Russian).

[83] S. L. Gorelov and M. N. Kogan, Solution of the linear problems of rarefied gas dynamics by Monte Carlo method, *AS USSR Izv., Mech. of Fl. Gas* (6) (1967) (in Russian).

[84] O. N. Korovkin and Y. I. Khlopkov, Solution of the problem of Knudsen layer with a slow condensation (evaporation) at the surface, *AS USSR Izv., Mech. of Fl. Gas* (4) (1974) (in Russian).

[85] Y. I. Khlopkov, Drag of a sphere in a flow of rarefied gas of slow speed, *TsAGI Scient. Notes* **6**(5) (1975) (in Russian).

[86] Y. I. Khlopkov, On the Brownian motion in rarefied gas, *AS USSR Dokl.* **222**(3) (1955) (in Russian).

[87] V. N. Gusev, M. N. Kogan and V. A. Perepukhov, On the similarity and application of the aerodynamical characteristics in transitional area by hypersonic speeds, *TsAGI Tech. Notes* **1**(1) (1970) (in Russian).

[88] Y. V. Nickolskii and Y. I. Khlopkov, Theoretical and experimental investigation of the low-density flow about a sphere taking into account evaporation and condensation from the surface, *TsAGI Tech. Notes* (5) (1980) (in Russian).

[89] Y. V. Nickolskii and Y. I. Khlopkov, The influence of evaporation (condesation) on the aerodynamical drug, *Proc., 15th All-Union Conf. on Actual Probl. in Phys. of Aerodisp. Media* (1989) (in Russian).

[90] A. A. Abramov and A. S. Kravchuk, Action of a thermal momentum on a surface in tangential flow, *AS USSR Izv., Mech. of Fl. Gas* (1), 139 (1994) (in Russian).

[91] B. N. Chetverushkin, *Kinetically Coordinated Difference Schemes in Gas Dynamics* (M., Publ. House of MSU, 1999) (in Russian).

[92]   M. I. Volchinskaya, A. N. Pavlov and B. N. Chetverushkin, On one scheme of computation of gasdynamical equations (Preprint No 113, *Keldysh Inst. of Appl. Math.* AS USSR, 1983) (in Russian).

[93]   I. V. Abalkin and B. N. Chetverushkin, Kinetically coordinated difference schemes as a model for description of gasdynamical flows, *Math. Modeling* **8**(8), 17 (1996) (in Russian).

[94]   T. G. Elizarova and B. N. Chetverushkin, Kinetical algorithms for computation of the gasdynamical flows, *J. Comp. Math. Math. Phys.* **25**(10), 1526 (1985) (in Russian).

[95]   V. P. Kolgan, Application of the principle of minimal values of a derivative to the construction of finite-difference schemes for computation of the breakdown solutions in gas dynamics, *TsAGI Tech. Notes* **3**(6), 68 (1972) (in Russian).

[96]   V. G. Krupa, On the construction of difference schemes with increased order of accuracy for hyperbolical equations, *J. Comp. Math. Math. Phys.* **38**(1), 85 (1998) (in Russian).

[97]   O. M. Belotserkovskii, *Comput. Mathematics* (3) (1957) (in Russian).

[98]   A. A. Abramov, A. S. Kravchuk and V. V. Poddubnyi, Statistical modeling of the surface outflow of a gas into the counterrunning flow, *J. Comp. Math. Math. Phys.* **31**(12), 1849 (1991) (in Russian).

[99]   Y. I. Khlopkov and S. L. Gorelov, *Monte Carlo Methods and their Application in Mechanics and Aerodynamics* (Textbook, M., MPTI, 1989) (in Russian).

[100]   A. A. Abramov, Direct statistical modeling of steady gaseous flows by small Knudsen numbers, *AS USSR Izv., Mech. of Fl. Gas* (4), 138 (1986) (in Russian).

[101]   Y. I. Khlopkov and S. L. Gorelov, *Applications of the Methods of Statistical Modeling (Monte Carlo)* (Textbook, M., MPTI, 1994) (in Russian).

[102]   P. Y. Georgijevskii and V. A. Levin, Supersonic flow about a body by the heat addition before it, *Proc. of Math. Inst. AS USSR* **186**, 197 (1989) (in Russian).

[103]   P. Y. Georgijevskii and V. A. Levin, Supersonic flow about bodies in the presence of the outer sources of heat release, *J. Tech. Phys. Lett* **14**(8), 684 (1988) (in Russian).

[104]   O. M. Belotserkovskii (eds.), *Numerical Study of the Modern Problems in Gas Dynamics* (Coll. of Papers, M., Science, 1974) (in Russian).

[105]   G. G. Chernyi, *Gas Dynamics* (M., Science, 1988) (in Russian).

[106]   V. I. Artemjev, V. I. Bergelson *et al.*, The effect of "thermal needle" before the blunted body in supersonic flow *AS USSR Dokl.* **310**(1), 47 (1990) (in Russian).

[107]   Y. I. Khlopkov, *Statistical Modeling in Computational Aerodynamics* (M., "Azbuka-2000", 2006), p. 158 (in Russian).

[108]   S. M. Belotserkovskii, Computation of a flow about the wings of an arbitrary plan form in a wide range of angles of attack, *AS USSR Izv., Mech. of Fl. Gas* (4) (1968) (in Russian).

[109]   G. A. Pavlovets and A. S. Petrov, On one possible scheme of computation of the breakdown flow about bodies, *TsAGI Proc.* (1571) (1974) (in Russian).

[110]   A. S. Petrov, Method of computation for unsteady breakdown flow about planar bodies by the viscous incompressible fluid, *TsAGI Proc.* (1930) (1978) (in Russian).

[111]   A. S. Petrov, Method of computation of the breakdown flow about elliptical cylinders, *TsAGI Proc.* (1930) (1978) (in Russian).

[112] A. S. Petrov, On the substantiation of the computational scheme of the breakdown flow about planar bodies, *TsAGI Proc.* (1930) (1978) (in Russian).

[113] A. S. Petrov, Solution of the Cauchy problems for Navier–Stokes equations in the form by Helmholtz, *Num. Meth. of Mech. of Continuous Medium*, Novosibirsk, **11**(7) (1980) (in Russian).

[114] A. S. Petrov, Application of the theory of Markovian accidental processes to the solution of Navier–Stokes equations for incompressible fluid, *Survey Appl. Indus. Math.* **12**(2), 253–264 (2005) (in Russian).

[115] A. S. Petrov, On the initial and boundary conditions for Navier–Stokes equations in the form by Helmholtz, *TsAGI Scient. Notes* **8**(2) (1982) (in Russian).

[116] A. Einstein and M. Smoluchowski, *Brownian Motion* (M.-L., ONTI, 1936) (in Russian).

[117] A. N. Kolmogoroff, On the analytical methods in theory of probability, *Math. Sci. Uspekhi* (5) (1938) (in Russian).

[118] F. H. Harlow, Numerical method of particles in cells for the problems of hydro-dynamics, *Coll. of Papres "Num. Meth. in Hydrodynamics"* (M., Mir, 1967) (in Russian).

[119] C. V. Gardiner, *Stochastical Methods in Natural Sciences* (M., Mir, 1986) (in Russian).

[120] V. V. Sychev, On the outsucking of boundary layer preventing the breakdown, *TsAGI Scient-Notes* **8** (1974) (in Russian).

[121] E. U. Repik and Y. P. Sosedko, Study of the intermitted structure of flow in the near-surface area of turbulent boundary layer, *Turbul. Flows* (M., Science, 1974) (in Russian).

[122] V. S. Sadovskii, N. P. Sinitsyna and G. I. Taganov, Numerical study of a mathematical model of near-surface flow within the turbulent boundary layer, *Near-Surface Turb. Flows*, Part1, Novosibirsk, Sib. Dep. of AS USSR (1975) (in Russian).

[123] Y. I. Khlopkov, V. A. Zharov and S. L. Gorelov, *Coherent Structures in the Boundary Layer* (M., MPTI, 2002) p. 267 (in Russian).

[124] V. E. Zakharov, The weak turbulence in the media with collapsoidal spectrum, *Appl. Math. Tech. Phys.* (4), 3–39 (1965) (in Russian).

[125] V. E. Zakharov and V. S. Lvov, On the statistical description of nonlinear wave fields, *High School Izv., Radiophysics* **43**(10), 1470–1487 (1975) (in Russian).

[126] V. V. Struminskii, On the possibility of application of the dynamical methods for the description of turbulent flows, *Turb. Flows* (M., Science, 1974) (in Russian).

[127] S. L. Gorelov, V. A. Zharov and Y. I. Khlopkov, Kinetic approaches of description of the turbulence, *Proc. of the 20th Symp. on Raref. Gas Dyn.*, Beijing (1997).

[128] I. G. Dodonov, V. A. Zharov and Y. I. Khlopkov, Resonance properties of the laminar and turbulent boundary layers, *Num. Modeling in Probl. of Aerodynamics and Ecology* (M., MPTI, 1998), pp. 19–31 (in Russian).

[129] I. G. Dodonov, V. A. Zharov and Y. I. Khlopkov, The localized coherent structures in boundary layer, *Appl. Mech. Techn. Phys.* **41**(6), 60–68 (2000) (in Russian).

[130] I. G. Dodonov, V. A. Zharov, Y. I. Khlopkov and K. Y. Gusarova, Renormgroup methods of theoretical description of the turbulent motion of a fluid, *12th School-Seminar "Aerodynamics of Flying Apparatus"* (2002) (in Russian).

[131] Suini *et al.*, *Hydrodynamical Instabilities and Transition to Turbulence* (M., Mir, 1984), p. 344 (in Russian).

[132] A. S. Monin, P. Y. Polubarinova-Kochina and V. I. Khlebnikov, *Cosmology, Hydrodynamics, Turbulence: A.A. Fridman and Development of His Scientifical Legacy* (M., Science, 1989), p. 325 (in Russian).

[133] A. S. Monin and A. M. Yaglom, Statistical hydromechanics, *Theory of Turbulence* **2**, 742 (1996) (in Russian).

[134] D. V. Shirkov, Renormalizational group, the principle of invariance and functional self-similarity, *AS USSR Dokl.* **263**, 64–67 (1982) (in Russian).

[135] E. V. Teodorovich, Use of the method of renormalizational group, eds. Monin and Yaglom, *Statist. Hydromech. Theory of Turb.* **2**, 742 (1996) (in Russian).

[136] L. C. Adjemjan, N. V. Antonov and A. N. Vasiljev, Quantum-field renormalizational group in the theory of developed turbulence, *Phys. Sci. Uspekhi* **166**(12) 1257–1284 (1996) (in Russian).

[137] A. N. Vasiljev, *Quantum-Field Renormgroup in the Theory of Critical Behavior and Stochastic Dynamics* (Siberian Publ. PIAF, 1998), p. 774 (in Russian).

[138] F. Kline, Lectures on the Development of Mathematics in XIX Century (M., Science, 1989), Vol. 1, p. 459 (in Russian).

[139] V. R. Kuznetsov, V. A. Sabelnikov, *Turbulence and Burning* (M., Science, 1986), p. 287 (in Russian).

[140] O. M. Belotserkovskii, S. A. Ivanov and V. E. Yanitskii, Direct statistical modeling of some problems of turbulence, *J. Comp. Math. Math. Phys.* **38**(3), 489–503 (1998) (in Russian).

[141] B. L. Kader and A. M. Yaglom Similarity laws for the near-surface turbulent flows, *The Results of Science and Technique, Ser. Mech. of Fl. Gas* (1980), pp. 81–185 (in Russian).

[142] I. I. Vigdorovich, Similarity laws for distribution of velocities and components of tensor by Reynolds in the near-surface area of turbulent boundary layer with blow-in and blow-out, *AS Russia, Mech. of Fluid and Gas* (5), 78–89 (2002) (in Russian).

[143] Y. I. Khlopkov, V. A. Zharov and S. L. Gorelov, *Coherent Structures in the Turbulent Boundary Layer* (M., MPTI, 2002), p. 267 (in Russian).

[144] A. V. Boiko, G. R. Grek, A. V. Dovgal and V. V. Kozlov, The physical mechanisms of the transition to turbulence in the open flows, M.- Izhevsk, *Regular and Chaotical Dynamics* (Inst. of Comput. Research, 2006) (in Russian).

[145] H. Schlichting, *Theory of Boundary Layer* (M., Science, 1969) (in Russian).

[146] I. G. Dodonov, V. A. Zharov and Y. I. Khlopkov, The localized coherent structures in boundary layer, *Appl. Math. Techn. Phys.* **41**(6), 60–67 (2000) (in Russian).

[147] V. A. Zharov, Application of the discrete fourier transformation to the study of the dynamics of wave packets, *Appl. Math. Tech. Phys.* **45**(6), 31–37 (2004) (in Russian).

[148] V. A. Zharov, Phenomenological analysis of the interaction of outer turbulent flow with turbulent boundary layer on a plate, *TsAGI* Preprint (1993) (in Russian).

[149] Y. I. Khlopkov, V. A. Zharov and S. L. Gorelov, *Manual for the Computer Analytics* (M., MPTI, 2000), p. 118 (in Russian).

[150] A. S. Bukin, On one statistical model of a turbulence, *Proc. of the Conference "Problems of Creation of Prospective Aviational Engines*, Inst. CIAM, Moscow (2005) (in Russian).

[151] T. Tun, V. A. Zharov and Y. I. Khlopkov, Modeling of a turbulent transition in the boundary layer by the Monte Carlo method, *Proc. of the 50th Scien. Conf.*, MPTI, Part 4, pp. 40–41 (2007) (in Russian).

[152] I. V. Voronich and Y. Zei, Results of the hypersonic flow about the flying apparatus "Cliper", *Proc. of the 50th Scien. Conf.*, MPTI (2007) (in Russian).

[153] V. V. Voevodin, Technologies of the parallel programming, Message Passing Interface (MPI), http:/parallel.ru/vvv/ mpi.html (in Russian).

[154] MPI for the Beginners, http:/www.csa.ru:81/~il/mpi-tutor (in Russian).

[155] I. Evseev, Use of PVM, Introduction into Programming, http:/www.csa.ru:81/~il/pvm-tutor (in Russian).

[156] A. L. Lastovetskii, Programming of the parallel computations on the non-uniform computer networks in the language mpC, http://paralell.ru/tech/mpc/mpc-rus.html (in Russian).

[157] V. V. Voevodin, System of parallel programming Linda, http://parallel.ru/vvv/lec7.html (in Russian).

[158] T-System, http://www.ctc.msiu.ru/program/t-system/diploma/node21.html.

[159] A. G. Chefranov, Parallel programming, Taganrog, TRTU (2000), p. 113 (in Russian).

[160] V. P. Dymnikov and A. S. Gritsun, Modern problems of the mathematical theory of a climate, *AS Russia, Izv., Phys. of Atm. Ocean* **4**(3), 294–314 (2005) (in Russian).

[161] N. S. Bakhvalov, *Numerical Methods* (M., Science, 1973) (in Russian).

[162] G. I. Marchuk, *Methods of a Computational Mathematics* (M., Science, 1989) (in Russian).

[163] G. A. Mikhailov, *Some Problems in the Theory of Monte Carlo Methods* (Novosibirsk, Science, 1979) (in Russian).

[164] V. Ivannikov, C. Gaisarjan, M. Domrachev, V. Ech and N. Shtaltovnaja, Extension of the language Java through elaboration of the parallel programs with a distribution over data with the help of a library of classes DPJ, *Probl. of Cybernetics, Appl. of System Programming*, Issue 4, M., (1998), pp. 78–100 (in Russian).

[165] S. K. Godunov, A. V. Zabrodin, M. Y. Ivanov, A. N. Kraiko and G. P. Prokopov, *Numerical Solution of the Multidimensional Problems of Gas Dynamics* (M., Science, 1976), p. 400 (in Russian).

[166] V. V. Voevodin and V. V. Voevodin, *Parallel Computations* (Sanct-Peterburg, BHV-Peterburg, 2002), p. 608 (in Russian).

[167] O. M. Belotserkovskii and V. A. Gushchin (eds.) *Mathematical Modeling: Problems and Results* (M., Science, 2003), p. 478 (in Russian).

[168] V. A. Gushchin and V. N. Konshin, Unsteady flows of fluid with a breakdown and transition about the bodies of finite sizes, *Etudes on Turbulence* (M., Science, 1994), p. 259 (in Russian).

[169] A. B. Zhizhchenko and A. D. Izaak, Informational system math-Net.Ru. application of modern technologies in the scientific work of mathematician, *Math. Sci. Uspekhi* **62**(5), 107–132 (2007) (in Russian).

[170] A. B. Zhizhchenko, *Algebraic Geometry in the Works of Soviet Mathematicians* (M., LKI, 2007), p. 64 (in Russian).

[171] Y. I. Zhuravlev, Extremal problems arising by the substantiation of the euristical procedures *Probl. of Appl. Math. and Mech.* (M., Science, 1971), pp. 67–74 (in Russian).

[172] Y. I. Zhuravlev, On the algorithms of discernment with the representative sets (on the logical algorithms), *J. Comp. Math. Math. Phys.* **42**(9), 1425–1435 (2002) (in Russian).

[173] N. S. Bakhvalov, N. P. Zhidkov and G. M. Kobelkov, *Numerical Methods*, 3rd ed. (M., Binom, 2004), p. 636 (in Russian).

[174] V. P. Ivannikov, S. S. Gaisarjan, K. V. Antipin and V. V. Rubanov, The object-oriented surrounding providing the access to the relational SUBD, *Proc. of ISP, Russian AS* **2**, 89–114 (2001) (in Russian).

[175] K. M. Magomedov and A. S. Kholodov, *Net-Characteristical Numerical Methods* (M., Science, 1988), p. 288 (in Russian).

[176] V. A. Iljin and E. I. Moiseev, Optimization of the boundary control over the string oscillations, *Math. Sci. Uspekhi* **60**(6), 89–114 (2005) (in Russian).

[177] V. A. Iljin, M. A. Imashev and D. A. Slavnor, The procedure of restandartizations recurrent on the number of loops, *Math. Phys. (TMF)* **52**(2), 177–186 (1982) (in Russian).

[178] Y. P. Popov, V. V. Kolmychkov and O. S. Mazhorova, Analysis of the algorithms of solution of the three-dimensional Navier–Stokes equations in the natural variables, *Diff. Equations* **42**(7), 932–942 (2006) (in Russian).

[179] E. P. Velikhov, Y. P. Popov, *et al.*, The role of a large-scale turbulence in the redistribution of angular moment in the accreational disks, *Astronom. J.* **84**(2), 177–184 (2007) (in Russian).

[180] A. A. Samarskii and Y. P. Popov, Difference methods for the solution of problems of gasdynamics, 4th Ed. (Editorial URSS, 2004), p. 424 (in Russian).

[181] P. S. Krasnshchekov and A. A. Petrov, *The Principles of the Construction of Models* (M., FAZIS, 2000), p. 424 (in Russian).

[182] Y. G. Evtushenko, G. M. Mikhailov and M. A. Kopytiov, History of the native computational technique and academician A. A. Dorodnitsyn, *Informat. Technologies and Comput. Syst.* (1), 3–12 (2001) (in Russian).

[183] A. I. Golikov, Y. G. Evtushenko and N. Mollaverdi, Application of the method by Newton to a solution of the problems of linear programming of the high dimensionality, *J. Comp. Math. Math. Phys.* **44**(9), 1564–1573 (2004) (in Russian).

[184] G. I. Arkhipov, V. A. Sadovnichij and V. N. Chubarikov, *Lectures on the Mathematical Analysis* (Textbook. M., High School, 1999), p. 695 (in Russian).

[185] V. A. Sadovnichij and V. E. Podolskii, Traces of the operators, *Math. Sci. Uspekhi* **61**(5), 89–156 (2006) (in Russian).

[186] V. M. Balyk, D. P. Kostomarov, V. I. Kukulin and K. A. Shishaev, Self-organizational approach to the construction of a variational basis, *Math. Modeling* **14**(10), 43–58 (2002) (in Russian).

[187] N. N. Kalitkin and D. P. Kostomarov, Mathematical models in plasma physics (survey), *Math. Modeling* **18**(11), 67–94 (2006) (in Russian).

[188] D. P. Kostomarov, L. S. Koruchova and S. G. Manjelej, *Programming and Numerical Methods* (M., URSS, 2007), p. 224 (in Russian).

[189] E. E. Tyrtyshnikov, *Analysis With Matrices and Linear Algebra (Lecture Course)* (M., MGU, 2004–2007) (in Russian).

[190] K. V. Rudakov, On certain universal limitations for the algorithms of classification, *J. Comp. Math. Math. Phys.* **26**(11) (1986) (in Russian).

[191] V. V. Rusanov, The existence of a limitational profile of the type of a shock wave for TVD-schemes, Preprint, Keldysh *Inst. of Appl. Math.*, No. 177 (1986) (in Russian).

[192] A. V. Arutjunov, Perturbations of the extremal problems with limitations and the necessary conditions of optimality, *Ser. Math. Analysis* (M., VINITI, 1989), **27**, pp. 147–235 (in Russian).

[193] A. V. Arutjunov, Conditions of the extremum, *Abnormal and Degenerated Problems* (M., "Factorial", 1997), p. 256 (in Russian).

[194] Y. N. Pavlovskii, *Imitational Models and Systems (Course of Lectures)* (M., FAZIS, 1998), p. 122 (in Russian).

[195] V. L. Matrosov, *Synthesis of the Optimal Algorithms by the Algebraic Closings of the Models of a Discernment–Discernment, Classification, Forecast* (M., Science, 1989), pp. 149–176 (in Russian).

[196] I. I. Bavrin and V. L. Matrosov, *Higher Mathematics* (Vladospress, 2004), p. 400 (in Russian).

[197] D. P. Kostomarov and A. N. Tikhonov, *Introductory Lectures on the Applied Mathematics* (M., 1984) (in Russian).

[198] L. N. Koroljev, *Structures of the Electronic Computers and Their Mathematical Software*, 2nd Ed. (Textbook. M., Science, 1978) (in Russian).

[199] L. N. Koroljev and A. I. Mikov, Informatics, *Introduction into Computer Scienses* (Textbook. M., High School, 2003), p. 341 (in Russian).

[200] A. N. Konovalov, Adaptive gradiental methods for the linear problems, *Proc. of the Intl. Conf. on Comp. Math.* (Novosibirsk, 2002) (in Russian).

[201] G. I. Savin, *The System Modeling of the Complex Processes* (M., FAZIS, 2000), p. 288 (in Russian).

[202] Y. N. Pavlovskii, *Imitational Modeling* (M., FAZIS, 2008), p. 236 (in Russian).

[203] T. S. Akhromeeva, S. P. Kurdjumov, G. G. Malinetskii and A. A. Samarskii, *Structures and Chaos in Nonlinear Media* (M., Fizmalit, 2007), p. 488 (in Russian).

[204] V. I. Berdyshev, The best trajectory in the problem of navigation over the geophysical field, *J. Comp. Math. Math. Phys.* **42**(8), 1101–1108 (2002) (in Russian).

[205] V. I. Berdyshev, New nodel of a navigation of the autonomous apparatus over the field of heights and its fragment, *Rus. AS Dokl.* **390**(1), 1–4 (2003) (in Russian).

[206] V. V. Rusanov, The characteristics of the general equations of gasdynamics, *J. Comp. Math. Math. Phys.* **3**(3), 508–527 (2003) (in Russian).

[207] V. V. Rusanov, Approximation of the boundary conditions in difference schemes, *J. Comp. Math. Math. Phys.* **20**(6), 1483–1499 (1980) (in Russian).

[208] V. V. Rusanov, On the non-uniqueness of the solution of a problem of the flow about a cone at the angle of attack, *Diff. Equations* **19**(7), 1262–1271 (1983) (in Russian).

[209] V. V. Vasin, I. L. Prutkin and L. Y. Timerkhanova, On the restoration of a three-dimensional geological boundary through the gravitational data, *Phys. of Earth* (11) (1996) (in Russian).

[210] V. V. Vasin, On the convergence of the methods of gradient type for the nonlinear equations, *Rus. AS Dokl.* **359**(1), 7–9 (1998) (in Russian).

[211] Y. I. Zhuravlev and Y. A. Fljorov, *Discrete Mathematics — a Textbook* (M., MPTI, 1999) (in Russian).

[212] A. S. Alexeev, A. A. Alexeev, B. G. Mikhailenko, G. N. Erokhin, A. N. Kremlev and A. E. Osipov, Extension of the open packet "Seismic-Unix" for the registration of the effects of elasticity in seismic reconnaissance, *Regional Forum Siberian Industry of Informational Systems* (Novosibirsk, 2002) (in Russian).

[213] I. I. Eremin, *Theory of a Linear Optimization* (Ekaterinburg, Ural Dep. of RAS, 1999), p. 312 (in Russian).

[214] Y. I. Khlopkov and S. L. Gorelov, *Monte Carlo Mehods and Their Applications in Mechanics and Aerodynamics* (M., MPTI, 1989) (in Russian).

[215] Y. I. Khlopkov and S. L. Gorelov, *Applications of the Methods of Statistical Modeling* (M., MPTI, 1995) (in Russian).